The Spoken Word III: Recollections of Dryden's History; The Shuttle Years

edited by
Christian Gelzer

Library of Congress Cataloging-in-Publication Data

The spoken word III: recollections of Dryden history : the shuttle years / edited by Christian Gelzer.
    p. cm.
 "NASA SP-2011-4552."
 1. NASA Dryden Flight Research Center--History. 2. Astronautics--Research--California--Rogers Lake (Kern County)--History. 3. Oral history. 4. Aerospace engineers--United States--Interviews. 5. Astronauts--United States--Interviews. I. Peebles, Curtis.
TL568.N23S66 2007
629.1072'4--dc22

                                                                                    2007047436

Table of Contents

Acknowledgements

George Grimshaw endorsed this project from the start and made it possible; without his support it would not have happened. He and Shirley King began directing me to past and present employees who might tell their story. Steve Parcel of Dryden TV was always quick to accommodate the interviewees, something quite helpful when schedules sometimes posed challenges. The Dryden Photo Lab produced incredible imagery, as always, in the hunt for pictures that enhance these accounts. Erika Fedorko conducted a series of oral histories for Dryden History Office, many of which helped this book directly, as did Guy Noffsinger of NASA HQ, albeit for a separate project; he was kind enough to allow me to make use of his work. Rebecca Wright and Sandra Johnson of the Johnson Space Center history Office, which has the agency's premier oral history collection, conducted interviews they did not know I would use, and I thank them. Terri Lyon took on the task of transcribing many of the interviews at a time when she was already heavily burdened with other work, and like other elements in such a project, without her effort there would have been no book. As editor, Elizabeth Kissling poured over the interviews with care and patience, making the final product a delight to read; she did far more than was asked and I am indebted to her. Steve Lighthill laid out the book with a flair I have come to take for granted, which is a mistake; one day it won't be available and I shall deeply miss it.

Finally, my peer reviewers gave the manuscript an extraordinary and thoughtful read, stealing time from other obligations (one from a human spaceflight project) to help sharpen and polish this book. I thank them deeply. Any remaining errors of omission or commission are mine alone.

Christian Gelzer
NASA Dryden
Edwards, California

Prologue

This is the third in a series of oral histories about NASA's Dryden Flight Research Center. This volume focuses on the space shuttle program and thus spans nearly three decades. Since the creation of the center's history office in 1996, coincident with the 50th anniversary of the center itself, the office has collected oral histories from the center's workforce. The benefit is more than a publication such as this: it's the compilation of corporate memory, which is invaluable.

Volume three ostensibly covers the shuttle years—a period starting with the Approach and Landing Tests in 1977—but in fact the story begins earlier than that, with the first lifting bodies and the X-15s, both critical antecedents to the space shuttle. Because the shuttles were prepared for flight at, and launched from, the Kennedy Space Center, and the astronaut office is at the Johnson Space Center, it is easy to forget that other NASA centers were involved in the shuttle program, but they were. In fact, Dryden and the Air Force had been flying piloted rocket planes into the stratosphere from an enormous dry lakebed in the California High Desert before NASA existed as an agency. Eight of the twelve X-15 pilots earned astronaut wings in the 1960s flying from that same location, piloting the world's first reusable space plane in the process. Both Dryden and the Air Force were integral to the shuttle program from the beginning.

Here then, in their own words, are participants in the shuttle story.

Christian Gelzer
Chief Historian
Jacobs/TYBRIN
NASA Dryden Flight Research Center
2013

Glossary

534 and 508: Purge and cooling units attached by tubes to the back of the shuttle after landing

ALT: Approach and Landing Test

AP: Air Police

APU: Auxiliary Power Unit

DFRC: Dryden Flight Research Center

DoD: Department of Defense

FRC: Flight Research Center (now Dryden Flight Research Center)

JSC: Johnson Space Center

KSC: Kennedy Space Center

LaRC: Langley Research Center

MDD: Mate Demate Device

MIPR: Military Interdepartmental Purchase Request

MSBLS: Microwave Scanning Beam Landing System

NASA 25: Dryden's mobile command vehicle for shuttle landings (a modified motorhome)

OMS: Orbital Maneuvering System

OV: Orbiter Vehicle (OV-099, 101, 102, etc.)

PAPI: Precision Approach Path Indicator

PIO: Pilot Induced Oscillation

SCA: Shuttle Carrier Aircraft

SPORT: FAA air traffic control center at Edwards AFB

TBO: Time Before Overhaul

SLF: Shuttle Landing Facility at KSC

STA: Shuttle Training Aircraft

STS: Space Transport System

TACAN: Tactical Air Navigation

X-68: Former FAA designation for the NASA Shuttle Landing Facility at KSC

Johnny Armstrong

Johnny Armstrong first worked for the government at the Army's Redstone Arsenal, in Hunstville, AL, in 1953. He joined the USAF in 1956 and was assigned to the Flight Test Center at Edwards Air Force Base where he spent the balance of his career with a brief stint at NASA in Hunstville again. As a civilian at Edwards he worked on lifting bodies and hypersonic vehicles, including the X-15, space shuttle, and the X-51. He authored more than 20 papers. Armstrong retired in 2012.

Johnny Armstrong, interviewed by Guy T. Noffsinger, Shuttle Documentary Interviews, October 2010.

When John Manke climbed into the cockpit of the X-24B for the first flight he found a note taped to the intrument panel: "Honestly now, have you read the instructions?"

Noffsinger: Can you give me your name and your previous title?

Armstrong: I'm Johnny Armstrong. I worked for the Air Force Flight Test Center. I have been doing that for about 52 years. I was privileged to be able to work the X-15 and the lifting bodies with NASA Dryden, which was an obvious extension when the shuttle came along, to continue to work with them with our good working relationship we had all those years.

Noffsinger: What did you do?

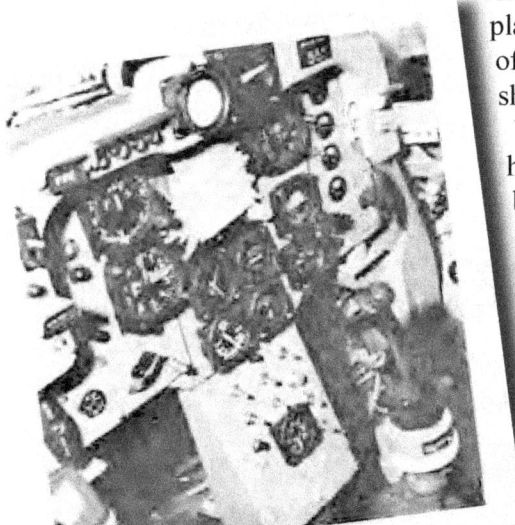

Armstrong: During the X-15 program I was one of the flight planners, as well as training pilots in the simulator. That kind of extended into the lifting body program, as well as into the shuttle program.

Very few people realize that the Air Force Flight Test Center had a shuttle simulation that was used for the early studies of both the Approach and Landing Test and the orbital flight test-landing phase. Joe Engle came here because we had worked with him on the X-15 and actually built the data maneuvers that he performed during the re-entry. You gotta remember that Joe was the only individual that ever flew the shuttle hands-on all the way from de-orbit burn through landing. In that flight plan he did data maneuvers by kicking the rudder and the ailerons and pitch axis to get the data, maneuvers that we later extracted to compare with the wind-tunnel predictions. He was the only guy to hand-fly it using cockpit gauges that told him when to do bank reversals and what angle of attack to fly all the way down through landing. He developed that on our simulator, then went to Houston. We had developed very close relationships with the engineers at Houston, who really weren't airplane testers but began to learn real fast, and we worked together with all the team members: Dryden, Johnson Space Center (JSC) and the Flight Test Center.

The Flight Test Center was interested because they were essentially hired by what's now called Space Command, but it was Space Division down in El Segundo [CA], because the Department of Defense (DoD) was going to have heavy payloads in that shuttle and so they hired us to [do] an independent evaluation of the space shuttle during the approach and landing.

Noffsinger: Tell me about the DoD, NASA, Air Force relationship?

Armstrong: That relationship was compelled by the DoD payload requirements, and once they formed that team we did, too. We worked as close team members all the way through and the same people were involved as on the X-15 and the lifting bodies: we were still the same team just trying to work to make something fly better, that happened to be called the space shuttle.

Noffsinger: Can you elaborate on lifting bodies, what they're used for?

Armstrong: Let me tell you how the lifting bodies impacted the space shuttle design. Early on, the space shuttle was going to have jet engines to return for horizontal land-ing, much like an airliner. The initial design was to put jet engines on the back of the tail, start them up and cruise in for a normal airliner type of operation. The lifting bodies had a significant impact: [they] plus the X-15, had proven specifically that they could make horizontal landings very accurately, unpowered, flying a steep glideslope, like 18 to 20 degrees glideslope. A normal landing comes in at about 3 degrees in an airliner. The lifting body program under [Milton O.] "Milt" Thompson presented a special meeting out here at NASA Dryden on those results to help convince JSC that their design did not need to carry that penalty of weight up to orbit; they could do a horizontal unpowered landing with the shuttle. That was the specific impact of both the combination of the X-15 and the lifting bodies to the design and all the operational aspects to a shuttle landing.

NASA Dryden is about three miles to the north of Edwards main base and the Ed-wards runway. We worked together on many, many programs and we have an excellent relationship that exists even today. They've even formed an alliance that includes both of those organizations, as well as the Rocket Lab [Air Force Research Laboratory] that meet routinely to see how can we continue to help each other and save money in the process. The Flight Test Center runway--the Edwards Air Force Base runway–it's the 15,000-foot runway.

Noffsinger: Describe your role in regard to the space shuttle: do you have anything you can tell me more about, like testing, about *Enterprise*, construction of the orbiters, anything like that, or are you more focused on the X-15?

Armstrong: Early in the program, when the orbiter was in Palmdale getting prepared for its first flight, we all had things we worried about. What if things happened? About that time, down in Mississippi, where they were testing the rocket engines, the turbo pump was having problems, and having failures of the turbo pump could be catastroph-ic in the orbiter.[1]

Over in Palmdale they made their first effort of putting the tiles on to protect the aluminum from the heating that we're gonna get on re-entry, and some of those tiles, whether they put them on in day time or at night, the next morning some of those tiles were on the hangar floor, so that was real scary. We could lose some tiles while they

---

[1]Modern liquid rocket motors rely on turbo pumps to force fuel and oxidizer to a combustion chamber because, among other things, compressed gas is not powerful enough to supply the mixture to match combustion rates. Moreover, gas pressure drops during the burn, reducing the flow to the combustion chamber while a turbo pump can maintain constant pressure. The first X-1s did not have turbo pumps and that was a limiting factor in the aircraft's performance; subsequent generations of the series were equipped with turbo pumps to feed the engine chambers.

Columbia landing on lakebed. NASA ED06-0045

were on orbit and really have a problem with the heat coming back on re-entry. They finally found a way to fix that by first putting the glue on there beforehand, letting it dry, and then putting the final [layer of adhesive] on it. We had nightmares worrying about what you do in those cases. As a matter of fact, I think there was some evidence that there were some tiles lost on the first launch and through some classified arrangement they were able to tell us not to worry about it because they had looked at it in some way, shape or form.

Noffsinger: Can you tell me your best memory from the program?

Armstrong: That's easy: when the first shuttle landed out here and did such a good job. Joe Engle was flying chase and to hear his voice coming in on final, talking to John Young and "Crip," Robert Crippen in the airplane telling him that they were right on glideslope, coming in for a landing, and then have John Young do his "Young dance," as I call it; as he got out of the airplane, walked around it looking up at the condition of the vehicle, he was just so elated. It was such a magnificent flying machine to him at that time.

Noffsinger: Which orbiter was that?

Armstrong: It was *Columbia* that made the first flight.

Noffsinger: Nothing that was scaring you, keeping you up at night?

Armstrong: Yes, all of those kinds of things worried me at night about the thermal protection system. My job was the re-entry, so the boost worried me because missiles had shown a habit of blowing up. One of the things we're glad never happened is the abort while you're going into orbit and have to do that return to the landing site where you turn around and use your engine to accelerate back to the landing site and land. Fortunately, they never had to use that, but it was programmed in to guidance to do that. I called that RTLS, Return To Launch Site abort. Some of our team members also had an influence in defining some of those aborts, like the abort across the Atlantic and to Rota, Spain, or one of those things if they had a failure too soon. We worked a lot of those kinds of failures, what-ifs, what could you do if those things happened?

You could get a bad impression from the shuttle if you consider just the accidents. The accidents were primarily caused by something other than the orbiter. The orbiter, as I think John Young called it, is a magnificent flying machine. It's a good airplane. The problems that it has had have primarily been from the external tank or the SRB [Solid Rocket Booster] , things that got it into orbit that caused the damage before it was ready to come down. The program was exciting from start to finish, starting with *Enterprise* and the release from the 747. That was another one that just had you on the

RECORD NUMBER OF RESEARCH FLIGHTS FLOWN

Reportedly, Chicago's O'Hare Airport has more traffic than any other airport in this country. However, the NASA Flight Research Center gave it a run for the money late last week and took a backseat to no one in the wide range of aircraft being flown.

According to the Flight Operations Directorate records, almost 73 research flights were flown at the Center during the five-day flight period. This figure does not count normal support flights.

Aircraft used in the activities ranged from small general aviation aircraft used to probe wake vortices caused by a commercial airline's Boeing 727 also operating from here, to the F-111 TACT and an aborted first rocket-powered flight of the X-24B. A new pilot checkout in the Center's Fly-By-Wire aircraft was also made.

Chase aircraft varied from F-104 jets to twin-engine Gulfstreams. Flight observers included Apollo 16 astronaut Tom Mattingly and representatives of the Federal Aviation Administration.

Center maintenance officials estimate that over 60,000 gallons of fuel were used during the operations.

The flights occurred on Wednesday, Thursday, Friday and Saturday of last week and carried over through Monday of this week.

edge of your seat, that that was going to lift off right 'cause my good friend [Fitzhugh] "Fitz" Fulton was in the cockpit of the 747.

The technology that came from this was very, very thorough. The data has been analyzed, it has been presented in conferences. The Flight Test Center has published about 10 reports on the first four flights about the re-entry, in excruciating detail, of all systems, the flying characteristics, all of that was well documented by a small bunch of guys--probably less than a dozen that the Flight Test Center had working the engineering side of the house. It was done under the leadership of Bob Hoey, who also went through the X-15 and lifting body program. He was the leader of that group at that time.

Bob Baron

Bob Baron spent 40 years at Dryden, working on such diverse projects as the Lunar Landing Research Vehicle, the hypersonic X-15, the Viking Mars parachute system, and the X-38 Crew Return Vehicle, intended as the lifeboat for the International Space Station. He earned numerous awards over his career and left a deep impression on those around him.

Interviewers: Curtis Peebles (Dryden History Office), Alan Brown (Public Affairs), and John Nelson (Bakersfield Californian newspaper)
Date: December 27, 2001
Place: Dryden Flight Research Center, Edwards, CA

I was born in New York in '39, in the Bellevue Hospital, which you heard is in downtown Manhattan. In fact, now it's a mental institution I think. In those days--they were delivering babies in 1939.

In '57 I went to Youngstown College. Back in those days, for whatever reason, I majored in physics, minored in math. I graduated in '61.

President Kennedy took office in January, and then in May 1961 Alan Shepard, Jr. went downstream a little bit in one of the Mercury capsules. At that time I was reading *Aviation Week* about what was going on.

At that time we were blowing up rockets left and right. I mean if you watched the news–there goes a Vanguard, there goes a Redstone. It was like the Vanguards would just melt on the pad. They'd fire and they would just collapse. And here is this crazy man--and I always told everybody at that time, and I still do, "I think he was smoking some pretty heavy Columbia gold," [saying 'we'll go to the moon'].

I graduated from college and came out here to work at NASA-Dryden, June 26th. My sister had just gotten married and come back from Florida and said 'Oh, I didn't like it. It was really humid and hot.' And, of course, I knew about Cleveland, where the Glenn Research Center is located [formerly Lewis Research Center]. I was from Youngstown and didn't want to go to Cleveland. And finally it was California, and so Sally and I said, "Wow, that's got to be wild." We had never been past Illinois at the time. And so we of said, "Well, let's go out to California." So I sent my resume.

Somehow I ended up on old Route 66, in Baker, CA, and it was in early June. I called my wife Sally, it was at nine o'clock at night, and I remember calling her up saying: "You won't believe this. It's so brutally hot." I'm looking at this map and it tells me I'm going into the Great Mojave Desert. I said: "What the hell have I been through the last day or two?" I think I ended up in maybe Mojave and spent the night. And when I got up the next day it was like, holy crap--am I on the moon? I said, "Jack Kennedy sent me to the moon. What the hell am I doing here? I've made the moon."

[For perhaps half-a-year Barron worked on a project called Avocet, a classified program at the time. All the flights took place at the Naval Weapons Center, China Lake, on Fridays and weekends to minimize unwanted scrutiny.][2]

---

[2]Peter W Merlin "The Evolution of Remotely Piloted Vehicles," in *NASA's Contributions to Aeronautics, Vol 2: Flight Environment, Operations,Flight Test and Research,* ed. by Dr. Richard P. Halllion. (Washington, D.C. NASA, 2010) 520-526.

One day we were there on a Saturday or Sunday, and Ted Ayers, who was kind of our deputy director at that time, under Marty Knutson, called up and said, "Come on home: Marty and I think we've got another project we want you to get started."

So I came back here and said: "What's this new project?" There was talk about building some kind of an airplane to do shuttle tests, like shuttle tire tests or shuttle roll-on wheel tests, or something like that, and they said: "We think with your experience, why don't you take that on and try to make that project happen." So we started the [CV] 990 project, which was actually called the LSRA, the Landing System Research Airplane.

I think one of the astronauts landed at Kennedy and decided to come back to the runway centerline. He stepped on the pedals a little too hard and by the time they rolled to a stop at the very end of the shuttle runway one of his main tires had gone through about eight cords; nine cords and it was going to blow. Some of the wizards here and at Johnson said: "Well, let's get an airplane and see if we can't modify it to simulate shuttle landings," and the Convair 990--old General Dynamics airplane--came about. We looked at a bunch of airplanes like DC-8's, 707's, something that was cheap, something we could get our hands on easily; lo and behold, the Ames Research Center had a 990. So we said, "Well, why don't we use one of our NASA airplanes so it won't cost us anything." That airplane, the 990, had a center keel keeping the whole airplane together. So, very simply, we cut the keel. We put the Brooklyn Bridge on board the airplane, and that is no kidding. We beefed up the structure like crazy on the outside so we could put it back together after we cut it. And then we bought these big hydraulic cylinders from, I believe, Denmark or Holland--we had to cut holes in the top of the airplane to get these rams in there, and that's an incredible modification. What we could do is come in and land at 250 knots: right in the middle of the two main gears there was another gear sitting there. That was the orbiter tire.

Kitty Hawk To Lancaster - The Hard Way

If you listen to Aeronautical Project's Wen Painter, it only took him 43 hours and 40 minutes of flight time to fly his 1945 J 3 Cub from Kitty Hawk, North Carolina to Lancaster, California. Furthermore, it only cost $168.06 for the required 189 gallons of fuel.

The first Wright Brother's flight at Kitty Hawk, that occurred 75 years ago, probably cost more than that.

However, what Wen doesn't mention is that it took him over 10 days with 27 refueling stops at airports and one in a cow pasture. And to fly the 40 hours, it took a total of 10 days, not counting the commercial flights back and forth to Norman, Oklahoma.

During one period of high head winds, a couple in an automobile made better time.

Wonder how long it would have taken Orville and Wilbur?

We actually put this Brooklyn Bridge structure on the airplane to make sure we didn't break the airplane. With these tremendous hydraulic rams we could drive that center wheel down onto the runway surface. If things went really awry that shuttle gear could come down and you could basically lift the rest of the airplane up, and you'd be a unicycle, which would be bad news to say the least. So obviously a lot of safety went into making sure that didn't happen. It was an incredible modification, about two to three years to cut a big size commercial jet in half.

We probably could not do that program today, probably couldn't even get started. In 1988 we could, because we were still taking some risks then.

I remember one of the fellows, Boleslaw Szuwalski, Bill, actually jacked the airplane up and kept it all structurally sound to cut that keel, which went boom--snapped, but all the other structure stayed in place. Then we had to rebuild the structure because we'd cut the main structural component of that airplane in half. We totally rebuilt it and besides that, put this tremendous Brooklyn Bridge structure on top so it could absorb all the loads that we wanted to get.

The thing about the 990 was that it could land at the speeds of the shuttle--about 250 knots. We put the spoilers up on the 990, we'd come in as fast as we possibly could so we could land at about the shuttle weight so the test tire would see 250,000 pounds of weight. The reason we wanted to check that out was because the orbiter launches from

Florida, and if it has to make a Return To Launch Site, an Abort Once Around, it has to come back really, really heavy. All the fuel is on board. Let's say the satellites are on board or the Hubble Telescope, and they're going to be really heavy: the tires are going to take all kinds of abuse and they [the Kennedy Space Center] have crosswind limits. So they were afraid of what might happen. I believe the astronauts would never want to do that: I think that even if they had just the faintest chance to get to Africa or wherever, they'd go in there. Trying to turn that orbiter back to Kennedy would be very tough.

The previous data the shuttle folks had gotten was from the Langley test track, which is only about a thousand feet long. But I do want to congratulate our Langley colleagues: there was all kind of information about what they were doing on their test track. They would get an orbiter tire and launch it on a surface for about a thousand feet. They could never get the speed or the weight, but then they would try their best to extrapolate what load factors the tire would really see on the wheel from what they saw.[3]

We spent a lot of time at Langley learning how they did these things, and then we went to Goodrich and Goodyear and the Air Force at Wright-Patterson Air Force Base, where they had these dynamometers and they test tire wear and stuff like that. The thing is, you're running a tire on a steel drum so you're generating a lot more heat and friction than in an actual landing, so the results are not really real, even though you may be putting a lot of friction on it and a heavy load. When you rode the airstream you would cool the tire: in the dynamometer you're always going on that steel drum; it was hotter than normal. So we actually proved that the dynamometers that were used throughout the country or the Langley track were very accurate only at the very bottom of the speed and duration range.

We started a program figuring that we were going to blow the tire and then see if the wheel disintegrated. What kind of damage would it do to the 990, we wondered? So we covered the whole bottom of the 990 with thick armor plating steel. We actually emptied one or two of the fuel tanks in the middle just to make sure they had no gas in them. We were going to blow tires and see where the shrapnel went. The other big concern we had was, if they blew a tire landing on the lakebed, would the wheel dig into the sand and make the aircraft roll over. So we started real destructive testing.

Because we thought this project out so incredibly well, we were able to do all this kind of stuff with that wheel. When it came down we could rotate it and put different drag on it, so it was seeing all kind of crosswind force. We had ten video cameras all around this thing and all kinds of data on the airplane. The shuttle folks said the orbiter had a 15-knot crosswind capability. As it turned out, the first time we took the 990 down to Florida and landed on that runway and went through a shuttle scenario we set the tire to replicate a 15-knot crosswind. About halfway through the test the tire blew, and everyone said: "Holy ____. What have we got here? What is this?" They told us to come back, saying we'd screwed up, "make sure to check everything out again."

We re-calibrated all of our instrumentation--went to the loads labs, all this kind of stuff--and we said: "Everything on our airplane is fine. Let's go back and do some more testing with this Michelin tire." We did more testing down there, we actually spent 30 or 45 days down at the Kennedy Space Center, and the second time we brought the fellows from Rockwell, JSC, Langley, Michelin. They would change tires after every flight even though there was hardly any damage on those Michelin tires. The tires were three thousand bucks apiece, maybe five now. We said: "Our data is really accurate. You guys have a little bit of a falsehood here about this 15-knot crosswind capability. You'd better not really say that." Everybody started believing the data, and they finally

[3]The Langley Aircraft Landing Dynamics Facility has been used to test landing and ground handling problems of aircraft. The facility has an 1800 foot track down which tires can be run (guided by a rail) while various conditions can be simulated.

said: "Well, this doesn't look good. Now that we believe this airplane and the test data, can we get more out of it? We'd like to go to 20 knot crosswind capability." So we got some Michelin tires and we started running tests, and we actually got up to a 20-knot crosswind capability, although we were up to six plies gone on the tire.

Langley people had said: "We'd better grind the runway down." They built this runway in the '70's, I'm sure, for the orbital landings and they brought things in, basically like big grinding trucks that came down and actually ground down the first two thousand feet on both ends of the shuttle runway, runway 15. Gordon Fullerton, super pilot and astronaut, used to land the CV-990 at 250-knots, like a crash landing, damn near, and we finally concluded, with the help of Langley--Bob Daugherty, superstar over there— that [the] modified the runway was okay. We finally did our tests and we said: "If you want 20-knot capability--grind the whole thing down," which they did. Yet they had not gone up to 20-knots. Now, when you hear them getting ready for launch the crosswind threshold is 17, 18 [knots]. They're still somewhat concerned that the astronauts will be so shook

Shuttle Simulations Made By DFBW

Inflight simulations of portions of NASA's Space Shuttle Orbiter computer flight control system were made last week and this week here at the Center, using the specially modified F-8 Digital Fly-By Wire (DFBW) aircraft.

The purpose of the flights, which simulate the maneuvers the Orbiter will perform following separation from the 747 Shuttle Carrier Aircraft (SCA) this summer, is to gather data to support the Shuttle's backup flight control system software.

The DFBW flight control system is very similar to the primary and backup control system in the Shuttle. The conventional mechanical control systems in the aircraft have been replaced by lightweight wires and electronics to translate pilot signals to the aircraft control system. Three digital computers are used for primary control, and a three-channel analog system is available for emergency control if the digital system fails.

Both the F-8 and Orbiter use similar models of the same digital computer. Parts of the Orbiter computer's logic for detecting failures of sensors, such as rate gyros, accelerometers and pilot stick position, have been programmed on the F-8 DFBW computers.

The F-8 was flown along a flight path which follows a racetrack pattern with the simu-

lated separation of the 747/Orbiter occurring about eight miles to the right and parallel to the landing runway on the dry lakebed. After the separation point, the F-8 pilot performed a series of test maneuvers, similar to those which the Orbiter pilot will perform, to obtain data on the flight control system.

Data was recorded and analyzed after the flight to determine the effect of pilot maneuvers and turbulence on the computer's ability to detect failures and avoid giving false alarms.

The F-8 pilot then initiated the first of two 90-degree turns to the left which aligned the aircraft with the lakebed runway at an altitude of 6500 feet, about nine miles from the point where the Orbiter would touch down. A preflare maneuver was made at an altitude of 900 feet.

In the case of the Orbiter, the landing flare would be initiated at an altitude slightly less than 100 feet. The F-8 pilot, however, made a low pass at approximately

Continued on page 3

up about having to return right then--that they'll be doing too much jiggling around and screw up the tires. Even though we had given them a 20-knot capability they keep it at about 17 or 18. They will launch at 17 or 18 knots now at crosswind, but that's about all. They won't go up to 20.

Chris Nagy and I used to give papers to the tire community, the landing community, the brake community--the Society of Automotive Engineers. We finally got the dynamometer people and Bob Daugherty's people at Langley to know the limitations of their systems and what the 990 could really do. The most exciting thing happened at the end of that program: we said, "How about if we really tried doing the things that we wanted to do on this thing. We know we can blow tires: that's kind of easy--we could do it purposely if we want to. Let's go back to what would happen if we blew a tire here on the Edwards runway. What would happen to the wheel?"

We always landed on the runway going toward the lakebed: in case [the] 990 had any problems we wanted to make sure we had enough space to stop. Sure enough, we blew the tire intentionally and rolled on the rim of the wheel. There was some thread still on the rim, but the rim was sparking and the rubber caught fire. Before you knew it, there was a tremendous fireball back of the 990. We had pilot, co-pilot, and five engineers in the airplane, and if they could have seen the fire going out the back they would have jumped off the airplane. It turned out to be just a fireball from the air blowing on it, and the fire was pretty well contained right there by the rubber that was burning a little bit. But it was spectacular.

As soon as the airplane stopped, probably six fire trucks showed up. We had Air Force fire trucks everywhere. I said immediately: "Okay. Now we know that the wheel will lock, it will smooth out the wheel and just grind to a stop. There will be a flat spot generating some sparks. Now that we're so brave, why don't we try one on the lakebed," which was the last flight of the airplane.

So we did.

We put a new tire on, and another wheel, and everyone said: "It's going to jam, it's going to get stuck." But we landed on the lakebed we blew the tire as we wanted to; the rim just kept going, going, kept on going, kept on going. When we actually stopped the test and went out there and looked at it, it was almost like someone had just taken a little bit of sandpaper to the rim. It was almost like you could use that rim again. You never would, but the data was that you didn't have to worry. You could land an orbiter on the lakebed here or at the White Sands Missile Range in New Mexico, and you were not going to break it up and dig into the ground. It was just going to keep on rolling and just grind down. You'd sandpaper all of it.

Roxanah Yancey, one of the original 13 employees to come to the Flight Research Center, passed away last Sunday after a long illness. Roxanah came to the Center in 1946, and had worked with nearly every experimental aircraft flown here, first in computation work and then in flight analysis. She retired last summer.

Two incredibly important results came at the very end: You'd roll on the rim on a concrete surface like Kennedy or here, and on the lakebed you'd just grind it down a little bit; no big thing. I think we had really, really impressive results.

The Air Force didn't want the program interfering with normal activities, so we came in on a Saturday. They said: "As soon as you're done we have some B-1's that want to do some flying." We were sending our 990 tires to be inspected and replaced or repaired to the Air Force. Somewhere along the line at their tire shop somebody left a couple bolts inside of a tire. And we were flying this tire that day . . .

Q: With the bolts inside?

Baron: Inside the tire on the 990. Well, it had been bouncing enough that a bolt

The Convair CV-990 rolling across Rogers Dry Lake during a shuttle tire test. NASA EC95-43230-4

This angle shows the test rig with the shuttle tire partially extended, as well as additional strengthening of the bottom of the aircraft's fuselage to guard against damage should the tire come apart catastrophically. NASA EC93-41018-18

actually blew a hole. So we stopped the airplane, but this one tire went poof. It just kind of went. Then the tire next to it went: you could see the tire next say, "Oh, crap. I'm going to carry too much load." The other seven tires, which were pretty warm, gave out. Within about five or ten minutes you heard pop, pop--and we were still on the runway! All eight tires on the 990 poofed. It was like their pressure plugs just gave way because the pressure went too high.

We had to call up the colonel, and he said: "Get your airplane off the runway. We've got some B-1's coming back!" But we can't--we're stuck. We had to go to Dryden, get spares, and come all the way back and change tires. We tied up that runway for four or five hours and Monday morning the general came in and heard this. I think he went right to Gordon Fullerton with a colonel and said: "Don't you guys ever do that again." We got our butts kicked on that test

Two shuttle tires showing different kinds of damage—and total failure—brought about by testing with the Landing Systems Research Aircraft. NASA EC95-43229-1 and NASA EC95-43211-7

Q: It took a long time to change eight of those tires.

Baron: Oh, yes. It took us like at least four or five hours. It turned out that we were doing some pretty rough test, and we probably ended where maybe a B-1 couldn't come in or something like that. All the fighter jets could still come in behind us and use the rest of the runway; it wasn't totally out of commission all day long.

Rick Brewer

Rick Brewer started working on aircraft in 1984, with Rockwell International's B-1 bomber. In 1993 he opened a Fixed Base Operator in Santa Maria, CA, where he worked with General Aviation aircraft for two years. He moved on to Heavy Maintenance Overhauls on commercial airliners for Maintenance Repair Stations and then on to Quality Control for several years. He was asked to take over the team lead for NASA's Shuttle Carrier Aircraft (SCA) in 2009, and at the time of the interview working the SCA/Shuttle Program to completion at the end of this year (2012).

Rick Brewer, interviewed by Christian Gelzer, NASA DFRC, April 11, 2011
Shuttle Documentary Interviews.

Gelzer: Today is April 11, 2011 and I'm talking with . . .

Brewer: Rick Brewer and I work for CSC, Inc. We contract to NASA, maintain all aviation services for NASA here at Dryden, across the country to El Paso and Houston. My job out here as team leader is to maintain the shuttle aircraft, the 747s shuttles for ferry missions after flights.

Gelzer: How long have you done this?

Brewer: Almost three years now.

Gelzer: What were you doing before you came here to do this?

Brewer: I actually was working in the Quality Assurance office here at NASA, bouncing around different projects on the aircraft side of the house with Lockheed and Northrop on some of their projects going on out here. I got a phone call one day from the site manager for CSC, Ken McDonald, who asked me if I would be interested in taking a job on the 747s. I inquired as to why: well, Pete Seidl, the old manager, retired.
  Pete's a wonderful guy, done his 30 years, had enough and wanted to go home and spend some time with the wife and on the golf course, and I was the only one apparently at the time that had enough wide-body [aircraft] experience, that would know enough about the position. So I went down and talked to Pete and talked to John Goleno, his right-hand guy, and went back and took the job right after about a 15-minute interview. So here I am.

Gelzer: What's your experience on heavies and wide-bodies?

Brewer: I've worked 747, 757, 767, heavy maintenance overhauls, heavy checks, all types of phases of maintenance on different Boeing airplanes. I've had my fingers in a lot of different areas which, when they said 747, was like "oh, it's another Boeing airplane: no problem;" stepped right in and pretty much took over. I have a crew of five and one inspector that work for me. We do all the work on the airplanes. We keep them airworthy and in flight status as far as ferry missions are concerned.

Gelzer: Talk about your experience with respect to the shuttle area in this particular job and the shuttles themselves.

Brewer: Well, it's been a lot of fun. For one thing, as a team lead/crew chief/supervisor/ site manager, I've been called a little bit of everything along the way. With only five guys working for you, one inspector that takes care of all your paperwork for you, life is good.
I have been fortunate that they've entrusted that responsibility, they let me run with it and keep the airplanes and decide what needs to be done, when it needs to be done and how it needs to be done maintenance-wise on the four-sevens.

   I've created a really, really great bond and friendship with the flight crews. They are based out of Houston, the flight engineers (FEs) and the pilots, great bunch of guys. I talk to them on a regular basis. They try to come out once a month and fly and stay current on the airplanes. I've had a couple new faces come into the program since I've been here but it's always a lot of fun. They guys are fun to work with, fun to fly with I have really had no bad experiences to speak of.

Gelzer: Are the crews from Johnson astronauts or are they strictly airplane drivers for NASA?

Brewer: The crews are strictly airplane drivers. They are dedicated. The chief pilot is a 747 pilot who has flown around the world for years and years and years on 47s. That is his forte, Jeff Moultrie. His number one man, Triple, we call him Triple Nickel, Jack retired a few months ago. We're trying to bring him back as a contractor. That would help us out through the last of the program for shuttle delivery and stuff.

   I have Bill Reike who is currently employed with Delta part-time and as a NASA pilot. Now, all the guys still fly NASA T-38s and stay current on the smaller airplanes. Bill is a

AN AMERICAN AIRLINES B-747, very similar to this Pan American 747 which visited the Flight Research Center several years ago, will be purchased by NASA for use in the Space Shuttle program.

747 TO BE USED FOR ORBITOR TRANSPORT

NASA plans to use a Boeing 747 to transport the Space Shuttle Orbiter and related Shuttle hardware cross-country.

The 747 will also be used in the planned approach and landing tests of the reusable Orbiter. The tests are scheduled to be held at the Flight Research Center.

This new concept replaces earlier plans to install six airbreathing engines on the delta-winged Orbiter for flight testing and for ferry flights from the West Coast to the Kennedy Space Center launch site in Florida.

A used 747-100 type aircraft will be acquired from American Airlines. Cost of the aircraft is estimated at $16 million. A team from the Flight Research Center will assist in the acceptance of the aircraft.

great wide-bodier. He knows 767s. He jumped in the 47 just like it was nothing. Two or three trips and he was ready to go, no problem whatsoever.

   Our other pilots out here: Bill Brockett has been around forever. Bill is a NASA pilot. He flies everything. The other pilot we have, Frank Batteas, with NASA, Frank obviously flies everything, F-18s, you name it. He's had some seat time in it. They are backup pilots. I've been on a couple missions with Bill. Frank hasn't flown on a ferry mission with us as yet.

Then we have our flight engineers. I have two stationed in Houston, Gary Ash, who is a quality manager for NASA at Houston proper. Another site manager from, I believe, El Paso. Steve Malarchick is site manager out there who fills in. He has been trying to get certified for two years but he just can't get enough seat time to get signed off. And Larry Larose also serves as an FE for us. It's kind of fun when he comes out because he is always working his books and practicing.

Then we have Henry Taylor, from Houston. Henry has been on the program for 30 years, knows the airplanes inside and out and is the Chief Flight Engineer. Henry and I are the ones that usually do most of the talking. We talk all the time. He says "I think this is wrong" or "What do you think about that" or "Should we put this in the maintenance schedule." We are constantly talking back and forth. It's a really great relationship.

Employees May Invite Guests To Flight

Several questions have been raised concerning guests at the Center during the free flight phase of the Shuttle Orbiter. The following information is presented to assist employees, both at Center headquarters and at North Base, who will be bringing their families out for these flights.

All family members should be pre-registered and badged in the lobby of the ISF prior to flight day. On flight day, guards will be stationed beyond the two main entrances to the Center and at the entrance to North Base, and only those guests with badges will be allowed to enter. Guests are encouraged to use one vehicle, if possible, to cut down on congestion in the parking areas.

Viewing area for the flight will be on the ramp between the main building (4800) and the Heat Facility (4820). No one will be allowed into this area until the 747/Orbiter combination has passed on its way to the main runway. After the flight, guests will not be allowed in the hangar areas.

Employees are reminded that they are responsible for their guests and should be with them at all times.

Employees should use the 120th Street entrance to the base, as traffic on Rosamond Blvd should be extremely heavy. If employees do come Rosamond Blvd., they should stay in the right lane, as the left lanes will be directed to the public viewing area on the hill.

Gelzer: When flying across the country, do they always use just one crew, or do they actually add a second crew now and again?

Brewer: It's always been the one crew. We just don't have enough to make two crews, but everybody wants to fly so I'll run four or five or six crewmembers at a time. I'll usually have three flight engineers because everybody wants to get some seat time on the airplane. Any and all pilots, if they're available. If they're in country and can get on board they want to fly a ferry mission. Absolutely!

Gelzer: What's involved in keeping the airplanes airworthy?

Brewer: We've adapted a program written specifically for the 747s. We fly public, which means we are in a registered airplane, but as a NASA asset we don't necessarily have to conform to all the FAA standards. For example, if I have an inspection that is due based on regular maintenance schedules, I'll look at the paperwork and go, "We've only flown twice this past year on this airplane, so I don't see the need for us to tear the airplane apart to do all this due to lack of wear and tear." That being the common denominator for most of the maintenance, they've let me run with a lot of the suggestions as far as what needs or doesn't need to be done, and I present them back to NASA, which decides in writing: okay we'll postpone this for 18 months. They base their maintenance on commercial schedule, which leaves us way out of the picture. We don't even come close to commercial standards as far as flying hours and landings.

Gelzer: Is the aircraft an experimental designation, given the modification to it?[4]

[4]NASA self-certfies its aircraft. It sometimes applies an "experimental" placard to FAA certified aircraft, such as the SCA, following modification.

Brewer: I think experimental is onboard the aircraft but it is a registered aircraft and it is designated as a public flying airplane. That allows us to put a crew roster together, i.e. employees or maintenance crew to fly with the airplane when it goes somewhere. NASA is very good about following procedures as far as requirements for the FAA.

Gelzer: Quite a treat! Do you recall any particular events of note in the years you've done this, things of interest, things that were out of the norm or unusual involving either the airplane itself or ferry flight?

Brewer: We've had a couple interesting ferry flights. I believe it was back in September of '09, we went back to the Cape when the shuttle landed out here at Edwards, and it landed very heavy. They brought down 4,000 pounds of trash from the ISS. That being said, when we loaded the shuttle on top of the '47, preparing it to depart. I had to sit down and reconfigure all the weight and balance sheets and fuel load capacities. The weight cut us a little short, shorter that normal. We had to make an extra fuel stop on the way back to the Cape. The max gross weight for the aircraft is 713,000 pounds; we left here at 710,000 pounds.

So, halfway to the Cape, in Columbus, Mississippi, they sterilized the airspace around the airport for us to land. In other words, no air traffic whatsoever, and we landed. I jumped off the airplane and the fuel trucks were waiting. Everybody was great, just "Johnny-on-the-spot" ready to go, and it was about 95 degrees and humidity was about 98%. It was hot. Our flight engineer at the time, Henry Marshall, who has since retired, came off the plane and I'm standing there in a puddle of sweat and he looks at me and says: "Is it hot out here or what?" I said: "Henry, based on the conditions I don't think we can stay on the ground very long. We have to go now." He looked at me and I looked at my thermometer, and it said 97 degrees in the sun. He ran back upstairs, called ATIS [Automatic Terminal Information Service], came back down and said: "Okay we're out of here." As soon as I disconnected the last fuel truck I looked at my guys, sent them back to the Pathfinder, and said: "Go, go, go! We're out of here!" They had the APU [auxiliary power unit] up and running on the airplane, the crew was back on board, we took off usually 20 or 30 minutes ahead of the '47; they took off first this time and used about 11,999' of a 12,000' runway. It was a very heavy day that day, almost didn't get off the ground. Triple was flying, Jeff Moultrie was in the right seat and they were giggling about it all the way back to the Cape.

We left there that same trip at about 11:00 a.m. or 12 o'clock their time and as we entered Florida airspace we hit thunderstorms, very bad thunderstorms. We were of course in the DC-9 Pathfinder cutting a path on the weather, telling the 747 where to go like we always do. Our weather radar went out and we didn't know where they were. We talked to them, they'd give us a location but that didn't mean anything to anybody on board the Pathfinder. We strapped in and soon we're flying like we were in the back of an F-18: we were punching clouds and looking for a space. Meanwhile, the 747 just lifted up, went left, and flew all the way around everything. We landed at the Cape, and as soon as I stepped off the airplane they were touching down right behind us. I'd never seen that happen before. It was a fun trip, it was a real fun trip. The weather is pretty unpredictable sometimes back there.

Gelzer: Once you're there and the shuttle has been unloaded, do you fly back on the SCA?

Brewer: Yes. The past few years, due to legalities or whatever, the administration has felt that they didn't need to have anything except necessary personnel on board in case there is an incident. They don't want to lose the crews and the assets so they won't let us fly on them anymore with the shuttles on board, but when we come home after everything is said and done, we preflight the airplane, throw our luggage back on board and we fly home on the '47.

Gelzer: Even without the orbiter there's a speed limit on the airplane?

Brewer: Yes, the airplane is designed to fly at 250 knots max. The aural warning system is set up to tell you: "Okay 252, slow down, back off the throttles;" horns and whistles and bells go off. It gets a little noisy upstairs.

Gelzer: Is this because of the additions like the vertical stabilizers?

Brewer: Yes, the external structures. It's designed to take quite a beating. When they modified the airplanes they went up into the tail section and completely beefed up all the internal structures. They added a computer on board on the horizontal stabilizing system. Keep in mind, these airplanes are 40 years old, so they're all original fly-by-cable. We have push-pull controls and cables; there are no computers on board as in a fly-by-wire system. All the gauges are analog. There are little digital readouts for fuel, which is standard from the factory, and that's all the digital equipment on board.

The computer for the horizontal stabilizer monitors the bungee feel system: when you put inputs in for flight control purposes, they limit the travel when the shuttle's on board, so I engage the computer for a ferry flight with the shuttle on board. When we take the shuttle off, I disengage it and it flies like a normal airplane.

Gelzer: A touch of fly-by-wire on a very old airplane?

Brewer: Just a touch. That's about the only place they've cheated.

Gelzer: You carry or you call for parts? You carry some parts if you need to replace them on a ferry flight?

Brewer: I do. On a ferry flight the only part of the airplane that's used for carrying anything is the forward cargo pit. When the shuttle sits on top it makes the airplane tail heavy so I have 1,700 pounds permanently bolted to the forward cabin on the main deck in front of the first-class seats that we've left on board: that weight never leaves the airplane.

When we ferry, I put 7,000 pounds of pea gravel on the very forward side of the forward cargo pit, and behind that I'll carry a couple tires, a jack to jack the airplane with to change the tire (which we've had to do on occasion), and an extra brake assembly. I have a couple of big bins - you've seen the luggage containers they use? I carry two of those full of parts. All kinds of parts, anything breaks on the wing, actuators, valves, we carry a lot of parts.

Gelzer: Have you had to use these parts?

Brewer: Yes.

Gelzer: During the flight?

Brewer: We've never had an issue where we've had an in-flight emergency, where something has failed in flight. We've never had a condition arise of any major concern in any ferry mission that I know of, even before I got here. We keep them, I would just

Discovery, (STS-114, which landed at Edwards August 9, 2005) hangs slightly off the ground in the Mate-Demate-Device at Dryden during servicing before its return to KSC. While at Dryden thunderstorms swept the desert at night providing a spectacular backdrop. NASA EC05-0166-22

say, in tip-top shape.

The biggest problem we have, our biggest nemesis, is the fact that the airplanes just sit so long, and when an airplane sits, it's a slow death. So, you've got all your electronics down in E Bay, and if they're not being powered up and used on occasion, I call them ghosts, we start chasing ghosts because a gauge might stick. So I try to go out there, at minimum every couple weeks, and power up the airplanes and move flight control around, exercise everything as much as I can. It seems to help.

Gelzer: The flight you mentioned that had the two tons of garbage that they brought back, that was the summer during which the orbiter was here a little longer than normal. If I remember, that was when there was a rainstorm.

Brewer: Oh, it rained good that trip. Yes it did!

Gelzer: After it stopped raining, for the two days after that the sun came back out and heated the place up to well over 100 degrees, it started to smell like a garbage truck.

Brewer: Yeah, it was rather ripe. A lot of people don't realize that when the shuttle lands it's had eight people on board for two weeks, it's a little ripe inside, too. It's not real pleasant to smell. It doesn't smell like a bunch of daisies in a flowerpot when you open the door. Of course, it doesn't help having the lavatory sit right there by the front door as you step in. So, with all the trash on board they thought they sealed everything up; well, the humidity just intensified the problem. It went aromatic if you will. It was quite interesting.

We had it here at Edwards. They opened it up and said "oh yeah they brought some extra things back." Usually procedure says when the United Space Alliance folks get in there, and I've been on board a couple times: they don't touch anything that comes back from space, including the trash. It stays where it's at, it stays tied down, all their experiments, if it's a controlled experiment where temperatures are required we'll leave power on the shuttle from the MDD [Mate-DeMate-Device] and we'll attach it once we're on the airplane. I'll plug it in and power from the airplane and maintain that temperature environment for them.

Gelzer: Why would they bring garbage back?

Brewer: Well, I don't know. I just don't know. I've talked to some of the astronauts over the years and you get different stories. They do eject out of the port-a-potty. It goes overboard and becomes little burning stars as it re-enters the atmosphere, but the trash itself I do not know.

Gelzer: And the trash was from the space station?

Brewer: Yes, it was.

Gelzer: And that was in the cargo bay. Is that correct?

Brewer: Yes. On the last mission [STS-133] they took that last section up for the station, and my understanding was, primarily for storage purposes. If anyone watches the NASA channel on a regular basis they've probably seen the video where they give you the tour. They float you through and you see different sections, and one of their escape pods right there is just full of stuff! They're just using it as a storage locker. It was just full because they have no place to put everything. So this last section that went up was hopefully to clear some of the clutter up and give them some more working room.

Gelzer: What is the airplane most prone to needing attention on?

Brewer: Mechanically? We use a lot of wheels and tires. We go through tires on a regular basis. Once a month the pilots try to come up and stay current. They fly mandatorily at least one time prior to every mission to make sure everybody is current. So they'll go up and fly for two or three hours and do, depending on how many people are on board, I've seen as many as 12 touch-and-go's and six full-stop landings and takeoffs. It's hard on tires. We break a lot of tires.

   But as old as the airplanes are now, I'll have an air pump go bad or a hydraulic pressure switch quit working. I'll get sticky gauges. Gauges might stick, or an N1 gauge might stick up in the flight deck and we'll just be changing gauges out.[5] Sometimes I'll have a fuel quantity gauge that's not reading right. I think primarily based on just lack of use.

Gelzer: Do Dryden pilots fly it once a month if the Kennedy people aren't going to be there?

Brewer: No, no. That's an interesting topic and I don't have all the information. What I will say is that Houston thinks -- says -- the airplanes belong to them. Dryden thinks they belong to it because they're out here, but the airplanes don't get flown unless Jeff Moultrie, the chief pilot from Houston, comes out and flies the airplane. Dryden pilots do not fly the airplane without Houston's approval and there will be Houston personnel on board when it happens. So, I tend to believe what I hear. People are funny. There

---

[5] N1 refers to the speed of the first spool on the turbofan engine, indicating engine RPM.

have always been rumors of territorial rights over the airplanes but it's subsided quite a lot lately.

Gelzer: Have they both been airworthy all this time?[6]

Brewer: Oh, they're both airworthy right now [April 2011]. They've given me written extensions on engines. I have two engines on N905NA that have 30 hours left, number two and number three, which has been a very critical point. Everybody's watching.

N911NA's main landing gear was due a major overhaul last December. Due to lack of cycles and flight time they said, "Okay, we'll extend it." I have until May of '12 before that has to be addressed permanently. Before that happens, we'll take two engines off of 911, put them on 905 and we're gonna park 911. We'll have enough time on the engines, of 911, that if we want to take it somewhere and do something with it we'll have about 30 hours before Pratt & Whitney says you're done, park it.

Gelzer: And they have, unlike recommended TBOs, a fixed hour?

Brewer: Absolutely! We suspected engine damage a couple years ago, came out and did a borescope on numbers two and three and found some leading edge damage internally on the stators in the engines, and they said "Okay." Pratt & Whitney came out, analyzed everything and said "We'll give you 60 hours of life on the motors, flying time." That being said, we've burned up 30 of them flying, staying current, and putting the airplane back into service.

When I got here 905 had been sitting for three years and I wound up doing two engine changes, main landing gear changes, a lot of black boxes. NASA just parked it

Three generations of vehicles in a row: in front, Dryden's Mini Sniffer, a small UAV designed for high altitude research, in the middle is Dryden's "new" C-47, and in back, the Shuttle Carrier Aircraft with *Enterprise* on top. The extra large cherry picker was to enable the shuttle crew to get into the orbiter. NASA EC77-8143

[6]TBO: Time Before Overhaul.

New Gooney Bird Replaces Old

It sits on the ramp at Dryden after providing nearly 23 years of cargo transporting service. Neglected now, since arrival of its twin, the C-47 aircraft tagged Gooney Bird because of its awkward flight characteristics, has been officially retired from cargo transporting duties.

In its place has risen a new emulator of the albatross bird. But arrival of the new C-47 didn't come without work to be done on it.

Reminiscing on the old C-47, Chester Pacewitz, C-47 Crew Chief, says it provided lakebed landing checkouts during X-15 flight research. It also served as a towing aircraft for the first lifting body M2-F2 and Paragliders.

Recently acquired from NASA Lewis, the new Gooney Bird has undergone significant changes.

According to Pacewitz, "We had to install new floor boards, new rugs, and completely re-upholster the seats. Also we removed both wings for checkouts and replaced the rudder and fuel system."

Clarence Haley, Chief of Aircraft Maintenance Branch, says the new Gooney Bird will primarily be used as a passenger cargo transporter.

Just recently passing major checkouts and flight clearance, the new C-47 is equipped with radar, auto pilot, and flight director.

But why a name like Gooney Bird? Pacewitz puts it this way; "It's like the big albatross bird, slow and awkward, that tries unsuccessfully to make it off the ground. But unlike the bird, the C-47 is efficient and reliable and despite its awkwardness is able to fly".

and walked away from it for the time being and flew only 911. So, when Pete Seidl retired and John Goleno, who at the time was our Quality Assurance person, said "We've got to change this and this," the list just multiplied overnight. We had months and months of work to do. It took me about nine months to get that airplane back in the air.

Gelzer: Do the airplanes fly exactly the same with an orbiter on them?

Brewer: Yes they do. They both are identical airplanes. The only difference between the aircraft is one is called a short range, an SR model. So if you look at 905, on the upper deck, what used to be the first-class cabin where the bar was, you'll see there are only two windows up there; 911 has five windows, a little bigger first-class on top. That was the difference between the standard long-range and the short-range model.

The fuselages are the same, the wingspans are the same; 911 has a little bit heavier gear on it, almost a -200 model gear versus a -100, which was a Boeing thing when they bought the airplanes. 911 came from Japan Air Lines and 905 came from American Airlines. So, it depended on what the customer wanted when they bought the airplane. For the most part they are identical.

Gelzer: And the orbiters, despite their own characteristics, don't make the airplanes fly any different, do they?

Brewer: Nope, just a little heavier. I had the pleasure of speaking with Gordon Fullerton many times over the years and Gordon did the first free-flight test with *Enterprise*. He told me that prior to that, in the wind tunnels, they had the '47 model and sat the shuttle on top and found that without the tail cone on it, in flight for an extended period it would actually take the tail off. The turbulence coming off of the orbiter would rip the tail off. So the tail cone smoothed everything out. They put the vertical stabilizers out at the end of the elevators and everything worked just great, but they found that with too much input it would tend to porpoise. So they limited the [elevator] travel to about 50% and no matter how much control input they don't get more than 50% deflection in flight and it smoothed everything out. They say it flies heavy but it flies smooth and when I was on it I couldn't tell a difference. We just took a little longer to climb and a little slower coming in but it flies really well.

Gelzer: Fuel consumption while the orbiter is on it is always an interesting thing to talk about.

Brewer: Yep, it does burn a little bit more. Average burn for us on a local flight is about 30,000 pounds an hour. With the orbiter we'll burn 40. It takes an extra 10,000 pounds of fuel an hour to fly and with the orbiter on top I can only put about 120,000 pounds on board which accounts for hop-skipping across the country for fuel. If we didn't have to stop for fuel we'd go nonstop all the way.

Gelzer: You have about a three- to four-hour window, plus reserves, to make it to where you fuel, 45-minute fuel reserve.

Brewer: Pretty much, about three hours. Yeah, exactly, and whenever we plan a ferry mission across the States, if we say we are going to go to Albuquerque or we're going to go to El Paso, there's always a backup plan. There's always an alternate landing spot wherever we go. For whatever reason, if we can't get there, we'll get there instead. So we never have to worry about running out of gas.

Gelzer: When making an approach do they always sterilize the airspace around it?

Brewer: Pretty much. I think John once told me that being a national asset, they never kept the shuttle carrier aircraft together. There was always one in Houston and one out here. Couldn't keep them together because it was so important, but I look at it like when you get a new bicycle: when you get it you don't park it next to anything, you keep it and you keep it shining. After you have it a few years, well, it got rained on today but that's okay, and you tend to relax a little bit, and I think that's been NASA's approach the last few years. "Yeah, well, we can keep them together." It's easier to maintain them if they're both here, and when they parked 905 this last time they were together it just never came up. I asked, and they said, "Oh don't worry about it, it's okay."

Last year was the first time we flew both aircraft at the same time, same airspace, and first time in the entire program.

Gelzer: What's the strangest thing that you've encountered in the time that you've had this position with respect to the airplanes? Or funny?

Brewer: We've gotten on board a couple times with the aircraft and we'll fire the airplane up -- say it's been sitting for a month and I know I had a hard failure in a particular box for example, our inertial navigation systems, for example: I would fire them up and one just would not come on line. Number one would just not work. Okay, so I'll go downstairs but I don't have a spare, so I'll order a box the next week, and then we'll be back up and we'll have to power up the airplane again and, having not turned the switch off when the guys powered down (which is not an uncommon thing to happen in that particular system), it comes on and works perfectly. We look around wondering what's different here and we'll turn it back off and back on, update the nav data and it works fine. That's happened several times and I can't explain why.

Probably one of the best times we had was when Buzz Lightyear was on board.[7] I have never seen so much attention paid to a toy in my entire life. I was kind of envious

[7]Disney's character Buzz Lightyear returned from space on September 11, 2009, aboard space shuttle *Discovery* (STS-128) after 15 months aboard the International Space Station. While on the ISS, Buzz supported NASA's education outreach program – STEM (Science, Technology, Engineering and Mathematics) featuring in a series of fun educational online outreach programs.

at times. This little guy got more attention than anybody I've ever known, better than the president, I think. We got the whole crew around him on the seats in the 747, got our picture taken. Mary Ann Chevalier, from Kennedy's Public Relations Office, was in charge of Buzz. She carried him in a little satchel under her arm 24/7, never left her side. When she came out here I kind of gave her the little finger and said, "Come here." We took her out in front of the MDD, and there sat Buzz out on the ground. She was kind enough to send me a disc with the photos later but we couldn't tell anybody until after they were officially released. We behaved ourselves and waited, but everybody got their pictures of Buzz Lightyear. That was pretty cool and I understand he's in the Smithsonian now.

I talked to the astronauts on that crew and they said, "Oh yeah, we let Buzz float around the cabin with us for a while. He took notes for us a few times and recorded some of the data." They taped him in places at one point, went to one of the consoles and put his hand on the start button for something. It was hilarious! They had a lot of fun with him. That was probably one of the funniest things I think we've done since I've been here.

On one rare day both Boeing 747 SCAs were airborne together and a chase F-18 with a photographer managed to capture the event. The nearest aircraft, tail no. 905, was the first of the two 747s that NASA acquired. NASA ED11-0237-16

Gelzer: That's the equivalent of a Flat Stanley in many respects.

Brewer: Pretty much. Pretty much. He was just the right size. You could pretty much do just about anything you wanted with him. We sat him on the ground a couple times and we would step back: with the imagery he would look like he was 6 feet tall. So we would have people walking in front of the shuttle when it was on the ground before we lifted and we'd sit Buzz right there and people walking by, Buzz looked as big as they did. It was some great shots. It was a lot of fun!

Gelzer: While the orbiter is actually being lowered or getting prepped for lowering, you're still involved up until what point? I mean, you will get involved once the '47 is towed in to the Mate-Demate-Device for lowering, right?

Brewer: When they raise the shuttle up in the MDD they [the servicing crew from KSC] say, "Okay, we're ready for you guys." We'll bring the airplane in. I've got to put that nose on a little 6-inch square. I only have so much room to get it in there. I'm allowed a foot clearance through the structure with the airplane, so it takes us a few minutes. We get in. We tie the nose down. Once we tie the nose down we power down, set the parking brakes, I pull my guys off the airplane, and we walk away. It belongs to USA [United Space Alliance] now: they go to work and proceed to lower and attach the orbiter to the airplane. They normally take about an eight-hour shift to do it.

My last trip to the Cape, it had been a long day, and it always is a long day when we go back, I got to the hotel about 8 o'clock that night. Two hours later my phone rings. "You can come push it back. We've got it off the airplane." It took them four hours to lift it off. So we got back out there. We waited an hour, an hour and 15 minutes and we pushed the airplane back and they were done. They set it on the ground. They set a record for themselves.

Gelzer: How long does it regularly take them to unload the shuttle?

Brewer: About six to eight hours.

Gelzer: Almost the same amount of time it takes to put it on?

Brewer: On average. Yeah. It depends on the crew and who is there. We've got twenty pages of a checklist to go through and once I get there we all go on headsets and I call the shots. I direct the tug driver in with the airplane and their monitoring. I've got guys with ears on upstairs watching the entire process, every inch of the way so that I don't touch anything, so that I don't scratch anything. It's a slow procedure. Nobody gets in a hurry.

Gelzer: When they first had the Shuttle Carrier Aircraft out here in the shuttle program, even after the ALT, the airplanes carried mission markings and at some point the mission markings disappeared? Do you know anything about this?

Brewer: I have researched it. I have had requests from several people to put the mission markings back on them before we retire. A lot of people would like to see that. We lost

With the consent of JSC, in 2012 Dryden photographer Tony Landis created new mission markings and applied them to SCA 905. NASA ED12-0096-02

the original markings; they were put on before the aircraft got painted, before NASA put its colors on there. When they got the airplanes, it had the American Airlines stripe down the side for example. They just put the NASA logo on the tail and flew it. Since it's been repainted and cleaned, everything came off. They started from scratch. I have requested it. I don't know, we'll see what happens. We'll see. I've had several people volunteer to do it. "We'll do it overnight Rick. You'll come in to work one day it'll be done you'll never see it." I said, "Great idea, but I don't want to have to take it off the next day when they see it and don't like [it]," it so I'm kind of waiting for approval on that.[8]

Gelzer: And the airplanes still carry the "9" on the fuselage, so technically they still belong to somebody else despite what people here may think.

Brewer: Yep, yep.

Gelzer: No word on why they didn't go back on, apart from that?

Brewer: Nope. I've asked everybody I could talk to. Pete [Seidl] looked at me and shrugged his shoulders, I don't know.[7]

Gelzer: Do you know what they plan to do with 905 once the shuttles stop flying?

Brewer: They have decided to keep 905 flying and they are currently looking at other avenues. You know we carried the Boeing Phantom Ray out here. That mission went extremely well, much simpler than everyone anticipated, much smoother.
I received a phone call from Northrop, asking if they could use our airplane to carry their X-47 and I said, "You know I have no problem with that but it's really not my call." I directed them to Houston, which was still in the middle of the Phantom Ray, and they said, "Let's do this." Houston engineering stepped back and introduced Boeing and Northrop and said, "Talk." Currently, I have the platform for the Phantom Ray, X-45, down in building 4833. Northrop's airplane is a little bit bigger, still has tricycle gear, my personal opinion: they could put it on there and we could carry it anywhere we wanted to. Little more weight, the flying characteristics would be practically identical. Northrop doesn't want to truck it and put it on a boat get it back to Pax River [Naval Air Station Patuxent River] but they won't finish their flight-test program for probably another 12 months, from what I understand, so it's on hold.
I received some calls from a couple of our local engineers here in the Loads Lab. Someone I knew from CEV [Crew Exploration Vehicle] project asked me out of the blue, "How many Gs can your airplane pull?" I said, "Excuse me?" And he said, "How many Gs can you pull?" I said, "Probably 1/10th less than it takes to break a wing, I don't know for sure." So I contacted a friend up in Seattle, a performance engineer for 747s. Our airplane will do 2 ½ Gs. So they were tickled. "Oh that's great! That's great! How much can it carry, Rick?" I said: "How much do you want to carry?" He says,

[8]In mid 2012 Dryden reapplied mission markings to 905 in a condensed form, in much the same way as they were on the NB-52B 008. 911 had, by that time, been retired from service.

[9]In the 1990s Dryden and the Air Force participated in a series of flight tests with Kelly Space & Technology to explore the possibility of launching rockets not from the ground, but from 20,000 feet, horizontally, after being towed aloft behind an aircraft. Doing so would eliminate the dispropportionate amount of energy consumed in the first phase of lift-off and boost, thereby making access to space more affordable. The tests used a Convair F-106 Dart towed behind a C-141 Starlifter. See Tom Tucker, *The Eclipse Project*, (NASA: Washington, D.C., 2000), Monographs In Aerospace History # 23.

"About 50,000 pounds" and I said, "Let me clear that with our flight engineer first. The shuttle only weighs about 200,000 so I'll get back to you." Now he's laughing! He's a great kid. We talk a lot. He said they're talking about putting a rocket on top and launching it off the '47. I said, "I only have one condition: no matter how you do this, when you do this, I won't be on board. You are going to have to pay the pilot's hazardous duty pay and you're not allowed to light the rocket engine until it's completely off the airplane." Now he's laughing, "'cause if you burn my tail off you're going to upset a few people." It's the only airplane I got left. But they are definitely looking at this. People at Goddard [Space Flight Center], people at Stennis [Space Flight Center], Houston and out here.

Joe D'Agostino

Joe D'Agostino came to Dryden following service in the Air Force, where he worked in Security Forces. At NASA Dryden he played a central role in the prototype shuttle *Enterprise*'s transfer from Palmdale to Edwards Air Force base the following year. There, *Enterprise* began the Approach and Landing Tests (ALT). D'Agostino's role in the shuttle program grew thereafter, culminating with his final position as shuttle program manager for Dryden until his retirement.

————————

Shuttle Documentary 2nd Interview by Christian Gelzer, June 6, 2011, edited August 2011 and April 2012.

Gelzer: Tell me your name and a little bit of your background and about when you got out here and what you did?

Gelzer: Please introduce yourself and when you came out here.

Joe D: I'm Joe D'Agostino. I came out here in 1976. I was hired specifically for 13 months in shuttle security operations. That was it.

Gelzer: For?

Joe D: For the ALT program and preparing for shuttle landings. It was kind of different in that I had been trained in the Air Force, spent all my time in security, well not all my time in security, but most of my time in security, and with things that were more sensitive, so I was aware of many of the requirements, whether it be working with classified [things], dealing with physical security or some of the things we talk about in law enforcement area, and one of my favorite has always been more the technical side of the house, in electronics and some of the new capabilities for using television to reduce the workload on people who were the human element. I came to work specifically on ALT. Dryden was entirely different. Things had already started to change before I came on board.

Gelzer: Had the MDD been built?

Joe D: Yes. One of the key players was Stanley "Ski" Markey. Ski was instrumental in developing the overland route from Avenue E to the base, developing that stretch of unimproved roadway -- it had always been a little dirt path to the duck pond by 10th Street East and then intersects with the entrance of base.

Rockwell was given the overall responsibility of developing the plans and procedures for moving the orbiters from Palmdale to Dryden. Headquarters was really in charge of that move but they delegated it to JSC. JSC, in turn, let us do most of the NASA coordination because we knew all the players, knowing some of the officials at the Air Force made the task easy; we talked the same language. But Rockwell did all planning,

[10]The "strongback" was transferred to the California Science Center to enable it to move *Endeavour* from the Los Angeles airport to its museum in 2012, where the shuttle went on permanent display.

they arranged with the contractors, cargo carriers down in Long Beach. A unit was developed -- the "strongback" currently in Area A along with a set of wheels, I think there were 64, placed underneath to transport the orbiter to here.[10] They did the actual day-to-day planning, whether it was working with the local police, with city planners in Palmdale and Lancaster for permits, looking at signs, trees, lights; they constructed a model which was no more than a flatbed truck with some poles to look at distances.

Gelzer: Width and height?

Joe D: Width and height. We trimmed a lot of trees, moved several lights and one of the challenges I had in working with the Air Force was how to get by the new gate, now the west gate, because the overland route actually is at the very beginning of base property (10th Street East and Avenue E) and the gate is about 10 miles away. You traverse a lot of land before you get to it and we had a telephone pole right at the gate that sat at 82 feet from it. The wingspan of the orbiter was 81 feet, however, the base commander had that pole removed, taking no chances.

Gelzer: The orbiter was brought up 10th; wasn't it brought across the lakebed instead of out to the gate itself?

Joe D: No, the actual route was from Site 8; it does a little jog, the jog is right where the current lifting platform is, or the flying A platform we call it [the Orbiter Lifting Frame]. It jogs around back towards the runway; one of the taxiways, then enters from the Rockwell roadway entrance to Ave M; at Ave M it makes a left-hand turn to go to 10th Street East, now Challenger Way; it enters Challenger Way [10th Street East], right there, goes straight north to Avenue E. At Avenue E it enters Air Force property, down that constructed temporary roadway -- because the Air Force wouldn't let us put in asphalt; it had to be temporary. So Ski and company came up with crushed rocks in a couple of the areas because it was kind of soft. You have to remember there are duck ponds out there so water naturally flows that direction. The route then comes via that roadway, which is actually almost 3rd Street East as we would know it. It finally reaches the roadway, actually two separate lanes in either direction, where the guards are at the West Gate, makes a right-hand turn onto the roadway leading into the base, comes through the gate, goes all the way down to the first intersection where the flash-

Enterprise Flies Alone On First Free Flight

The first free flight of the Space Shuttle Orbiter Enterprise is slated for takeoff today from the Dryden Flight Research Center at 8 a.m. PDT. Astronauts Fred Haise and Gordon Fullerton will be at the controls of the Orbiter for the approximately five minute long free flight. Flight crew for the 747 Shuttle Carrier Aircraft (SCA) will be Fitz Fulton, Tom McMurtry, Vic Horton and Skip Guidry.

This initial solo flight follows a series of unmanned and manned captive test flights conducted at Dryden which began in mid-February. The Orbiter was carried aloft for a series of five "inert" flights (Orbiter systems inoperative) before astronauts Haise and Fullerton and fellow ALT crew members Joe Engle and Richard Truly flew subsequent captive flights.

The captive flights verified the aerodynamic and handling capabilities of the 747/Orbiter combination as well as Orbiter systems and crew procedures.

The flight path of the Orbiter and 747 follows a racetrack pattern with separation occuring when the vehicles are about 13 kilometers (8 miles) to the right and flying parallel to the landing runway. From the separation point, the Orbiter will fly a U-shaped ground track to the runway.

Touchdown airspeed is about 200 mph KEAS and elapsed time from separation to touchdown is about five minutes, 10 seconds.

The next Shuttle flight is tentatively scheduled for August 30.

ing light is, makes a right-hand turn to head back south because you're heading toward South Base, and there was an actual old entrance to the base that we sort of improved, and entered in from that little improved road and then to taxiway Alpha.

Gelzer: And then you just came right down?

Joe D: Then we came down to here. That was the actual route for the overland transportation. Initially, the first time I can remember timewise; we started at 4:30 in the morning at Plant 42. That was where we held our briefing, where Rockwell briefed us. One of my roles was to relay that information to the Air Force because they would meet us at the end of Avenue E, where Air Force property starts.

Gelzer: You came up the road with *Enterprise*?

Joe D: Yeah, it was kind of interesting because up until that point we had seen very little assistance from KSC and headquarters; occasionally I would call them and tell them what was going on. They showed up that day. I was sort of used to that; it didn't bother me. We had a few little minor incidents that we had to be really careful about. I say we; I'm talking about the team now, not specifically NASA, about Rockwell and all the players. Highway Patrol and the Sheriff's departments that assisted us, specifically Los Angeles County, were leapfrogging ahead to make sure the intersections were clear, and one or two cases we had to move the lights; physically had to be moved and rotated.

The most important thing to remember during those days was the reaction of people in that environment as opposed to today's environment, which is security. It was completely different then. I used to say the people in the Antelope Valley were special because they had built the orbiter, it was their pride, they had relatives, friends that had worked on the orbiter. It was one of the premier programs in the Valley and they all felt it was part theirs. Moving it was a very festive time. I saw things that I didn't expect to see: kids up on fences with flags waving, parents that actually let their children get up on some of the 6-foot block walls; kind of worried they might fall off but still, very enthusiastic, clapping, smiling, hoorahs, waving a flag. We didn't worry about terrorist activity; we were aware that we certainly didn't want anything to happen and were looking for people running across the route, because we had those -- people would cross the road in front of us. We'd be coming up the street and they wanted to get in front, people wanted to get pictures. It was a different era.

Gelzer: How long did it take?

Joe D: We started at 4:30 in the morning. We wound up here at around 3:30 to 4:00 in the afternoon; it was a long, long day. I think from start to finish, the first time was almost 13 hours. That got better. We reduced it quite a bit.

Gelzer: The next one was *Columbia*?

Joe D: The next one was *Columbia*. Again, the most important thing from my point of view: I had two responsibilities during that time. The first one was security; second, I was the transportation officer so I worked very closely with JSC transportation people about vehicle requirements.

The other interesting thing about the overland route: I remember seeing tiles fall off the shuttle.

Gelzer: And that's just the overland trip?

Joe D: That's just the overland trip. What people didn't know was that a lot of those tiles were fake. They weren't the real tiles.

Gelzer: This is still the *Enterprise*?

Joe D: This is still the *Enterprise*. But we did see some of that later on with the first flight vehicles.

Gelzer: You lost some in flight, too, with the ALT. I do know that.

Joe D: Yeah!

Gelzer: What do you remember of the Approach and Landing Tests?

Joe D: Quite a bit. Ken Iliff asked me if I remembered seeing some puffs of clouds during the second flight and I said yes. We were the only two people that remembered that, apparently.

The flight one; most of us had been to the briefings every day, going through what we call the 72 hours, 11-day schedule over what would happen if something occurred during separation. I mean, we had watched it, we had watched the practices, we had watched the taxi runs, we felt confident there would be no problem with liftoff, no problem with getting in to the separation phase. Still, from a non-technical point of view the concern was: during one of the tests [what if] the orbiter did hit the tail? What if it doesn't separate? But I was too busy doing crowd control. We had a parking area set up on the hill to watch ALT. We had bleachers. We had parking to worry about. We certainly had to worry about the security.

Gelzer: There's a picture of [Gordon] Fullerton and [Fred] Haise having breakfast before the flight and they're dressed in their blue flight suits. It's almost as if they are getting ready to go into space. They have a special breakfast, they're sitting at a table, they're being looked at by the press and they're just busy eating their breakfast.

Joe D: That was sort of the way things had gone at JSC and KSC during the whole program so it's one of those logical follow-ons. Pilots have unique ways of doing things.

Gelzer: Here's a question then that you may or may not have any insight into. Neither of the first two men who flew the shuttle flew any of the Approach and Landing flights.

Joe D: Except for Engle and Gordo.

Gelzer: The first two to actually go in STS-1 had never flown in an ALT. Why were they not picked to fly at least one of the Approach and Landing Tests so that they had some first-hand experience?

Joe D: I can't answer that. I really can't. All I can tell you is that I was always puzzled why Fred never flew in the program, because the other three did. Gordo, Dick Truly, and Joe Engle flew into space.[11]

Gelzer: Then you had the PIO?

---

[11]Haise resigned from NASA in June 1979, two years before the shuttle flew, becoming the vice-president of space programs for the Grumman Aerospace Corporation.

Joe D: The PIO was kind of different only because, from my security perspective, where people were located. We had this requirement during ALT going back to day one: nobody would go out to the orbiter, we wouldn't have anyone near the orbiter except the convoy.

That issue was critical because one of the first things we heard is that the general, who was [Tom] Stafford at the time, former astronaut, commander of Edwards Air Force Base was going to go out to the vehicle because those were his buddies, his group, and yet, from my perspective I had instructions that nobody would go out except the crew convoy. Ironically, we had this big meeting and "Deke" [Donald Slayton] stood up and asked Chief Willy Bell and I to meet with him afterward because we both had concerns; Willy from the fire perspective, me from the security point of view were saying that there were additional people on the lakebed and we didn't have the force to cope with them. Deke said, "Do you have any concerns?" I told him specifically about the additional people and he looked at Chief Bell and myself and said, "Don't worry about it, it's taken care, of there will be no one going out." Wouldn't you know it, it turned out I had the keys to the general's car the day of the landing.

Gelzer: So he wasn't going anywhere?

Joe D: He wasn't going out. Then here comes the last ALT flight and we're coming on to runway 04, the concrete runway. We were going to take the Prince [Charles, Prince of Wales] out to the lakebed at the distance of a couple hundred feet from the landing. It actually turned out about 300 feet at that. We had a red carpet out there, there's pictures, had glasses, I won't say what was in them but they had glasses out there and we're watching this landing. The Prince had flown in that morning, I believe, because he'd asked a question about some of the patches of green that he had noticed flying in. He had been a guest of Rockwell and we'd told him those were alfalfa fields.

The liftoff was nominal, nothing to speak of. Separation went well -- I could listen on a radio. It was nominal. You're focusing, trying to focus on two jobs. You want to be a spectator but at the same time you're caught up in your job, security in my case. And then, as the vehicle approaches it looks like it's a little hot. Then I see the first bobble, "That ain't in the program." The second turn actually occurs almost directly in front of the VIP area, and you're saying, "Oh, no," your mind is already preparing for the worst and you're hearing some of the conversations you're not supposed to hear. I don't know what the exact words were but I always remember, "Get off the stick, get off the stick," and then the vehicle levels out and goes down the runway. The Prince looks around and you know he knows something is wrong. Nobody knows the answer at that point and then we start talking about pilot-induced oscillation.

Gelzer: Who was turning the *Enterprise* around for another flight? KSC contractor personnel?

Joe D: KSC. KSC had their own convoy commander, starting with ALT, even to the operational phase. They did all that. Dryden was really in a support role. We were gathering information from the Air Force, passing information on, providing things that they needed. We did have equipment.

We learned a lot from ALT in terms of how to do things such as crowd control and badging. Roger Barnicki came up with buttons for badges at one point. One of the

Second Shuttle/747 Flight A Success

The second manned flight of the Space Shuttle Orbiter "Enterprise" has been successfully completed. The Enterprise took off at 7:50 Tuesday morning with astronauts Joe Engle and Richard Truly riding the piggybacked orbiter on a 62 minute flight.

things we didn't want to do – I didn't want to do – was issue badges to every visitor. You can imagine how work-intensive that would have been for us since we had no setup to do this. My philosophy was that if you were a NASA employee and you said they were your guest, you let them in.

The OPS [Operations] world really changed as the years went on. Let me define support: support was anything they needed [in the shuttle program]. Literally, if they needed something, whether it be from local contractors, in-house or even the Air Force, it came through that organization. It sort of was a clearing-house, and it had a pretty good management philosophy in how they operated it. You must also remember in those days we didn't have things like tire shops for the aircraft. Tire shops for the shuttle, specifically the buildup of tires even during the ALT, was done not by Rockwell, but by the Air Force, which was kind of different. Most people don't remember that, but they did all that.

Along the way you had the rise of Dryden in taking on some roles that were not thought capable of by Dryden people. Some of the key players were Ed Sabo and Charlie Baker. Charlie had a lot to do with setting up some of the operational requirements in terms of the microwave scan beam landing system (MSBLS), the landing aids -- the PAPI (Precision Approach Path Indicator) lights. Those were later modified, but he did a lot of the work with JSC specifically, and KSC.

One of Dryden's [big] contributions came during STS-3, which really was a major go/no go decision made at 4 o'clock in the morning. We gathered, the weather looked bad at Edwards; lo and behold, less than four hours later we were getting ready with trains and everything else to do a relocation operation move, which was quite extensive. [To load all the ground service equipment onto trains and move it to the White Sands Missile Range in New Mexico.] It showed what happens when you take great planning, operational requirements, a project manager that literally was at KSC when the decision was made, and how Dryden played a key role in getting that train ready, because everybody else left us literally after things got going. They had to be at White Sands. So the White Sands move is important not only from a program viewpoint, it provided the development of Dryden operational and support techniques.

Photo and TV were just two examples of those lessons learned. Television was really the brainchild of a JSC employee, Fred Koons. Fred really was the most important driver in setting up the television program and understanding the importance of television in today's environment. It was only through his efforts that we developed what we have today. We took that experience, added to it, and played a key role in developing television requirements. Not as full-fledged as we have today but the fundamentals.

Gene Edmonds, another JSC employee, another of the driving forces that led to photo and the television plan. This was important because Dryden was a very small organization: its photographers weren't used to large-scale operations and here you have this grand-scale program that required new and different equipment. In the past we had usually looked for the Air Force for getting that help or equipment.

One of my shuttle highlights was designation as a "photographer." I'm listed as one of the official STS-3 photographers. The reason for that was quite simple: it was a favor that Gene gave to me. He gave me a camera -- threw it at me -- and said, "Take all the pictures you want." At the end of the flight he gave me the film only to discover later that all the other 35mm cameras had failed. I had the only official landing photos by a NASA person, which was kind of ironic. Funny from my point of view because one of the first things I did after getting my film back was have it developed by the Army. The Army developed pictures because I needed the pictures to help get more things that we needed, trading material.

Gelzer: This is the Army at White Sands?

Joe D: Yes, at White Sands. I had this Nikon camera that I took 36 photos with, but never worked a high-end camera in my life, had 36 pictures all in frames except when the shuttle [was] exactly in front of me and was too big to capture. I had all of the deviations in the flight; they had wind concerns and the orbiter was kind of bobbing and my photos showed the whole sequence, all the way to the ground. Don't ask me how I did it. Gene made a special picture, gave it to me after he had enhanced it of course, but that was one of the highlights for me personally.

Challenger incident, accident is important in terms of knowing the crews, having been with the families here at Edwards and also at JSC. Having been there, having worked with them on a daily basis, built some bonds up. The road to recovery [Return To Flight] was different because it gave me an opportunity to be away from the shuttle program. I actually worked in flight ops as an assistant scheduler for over a year, so I got away from it and looked at it from a different point of view.

I think Dryden played an aggressive role in improving the shuttle program, essentially saying, "Since we have this new capability [an improved MSBLS] we need to look at different ways of support." We started to develop processes that nobody envisioned. For example, we promoted the use of the opposite end of a runway, we encouraged the use of lakebed runway 15 for training flights -- which I don't think we'd ever used for landing. Why was that important? Because it was the first landing in which we had no real advance notice, literally one hour's warning that they going to land here. The commander at the time was Bob Crippen. Having spent many hours with that crew in training, I knew he had always hinted that he liked runway 15; he was sort of biased toward it, and lo and behold, at 7:30 in morning, when people came down to Dryden, there sat the orbiter in the middle of runway 15 [STS-7, OV-99, June 24, 1983].

As the program developed there were concerns. The lakebeds were pretty well defined in terms of hardness. We were very comfortable with using the lakebeds but there was some skepticism about the tow-back areas because they hadn't been used that much. You quickly moved to the concrete or asphalt, but along the way you had to cross certain areas of the lakebed that really hadn't been tested for all-up weight.

Certainly one of the key elements involving the Air Force from a historical perspective was crash and rescue. They were very important to the program because they were responsible for rescue, and for disaster preparedness. Their fire and rescue people were key players not only through ALT, but through the whole development phase, especially the fire people, who had to go into the orbiter and get familiar with systems. Unfortunately, the person who developed a lot of the systems that were responsible for that passed away a couple years ago and that's Fire Chief Willy Bell.

You had two super organizations (JSC & KSC) directed by an outstanding JSC project manager, "Deke". This was unique because one of the things he did was providing leverage for Dryden in a role that people often overlook: financial management. He promoted JSC keeping control of shuttle money -- that money being transferred directly to KSC. Why is that important? Because that enabled Dryden to have its own budget even when there were more and more landings at KSC. Having its own budget enabled us to do many other things that are in place today.

One of the most important as far as I'm concerned was the infrastructure upgrading. What you see today and what you saw at the beginning of ALT are so different.

Joe D: STS-7 was important to John [McKay]. John got his first indoctrination into the shuttle when he opened the hatch during STS-7, which was the Bob Crippen landing on runway 15 because he was the one that went out to open the hatch. People forget what it's like when you've been in orbit for several days

Gelzer: It's a smelly place.

Columbia touching down on Northrup Strip at the U.S. Army's White Sands Missile Range, NM, March 30, 1982. STS-3 was the only shuttle flight to land at any site other than KSC or Edwards/Dryden. STS-3 Photo courtesy of Joe D'Agostino

Joe D: The space develops its own environment.

Gelzer: I was interested in your memories of pulling the orbiter off the lakebed early on and the wheels getting mired.

Joe D: Stuck twice.

Gelzer: Stuck twice?

Joe D: Yeah, I'm partly to blame for one and partly take credit for one. The landing took place early in the program out on lakebed runway 23. We had that transition to make from the lakebed to the Dryden ramp, and one of the early decisions was that we would come directly back to the MDD.

One of the things we had learned was the orbiter was getting heavier and heavier, going from 180,000 to well over 220,000 some-odd thousand [pounds]. So we are on the way, towing back. We had picked a route we felt very comfortable with. As we got a couple hundred feet away from the ramp, the vehicle started to plow. What do I mean by plowing? As you move more and more into what [are] sand-like conditions it starts going a little deeper, a little deeper, it gets harder to pull. It got to the point where the alarm went off. In the tow vehicle there are a couple safety devices that tell you when you're exceeding the amount it can pull; a flag pops up and a horn that goes off, and actually there are some mounts inside of it, if I remember they're rubber mounts that separate at certain points so you can't get into a situation where you overstress [the nose gear on the orbiter]. The nose wheel started to go down, literally started to sink a little deeper, and a little deeper -- we might have made it but you couldn't take the chance. There was quite a debate about what to do and one of the first things that was suggested was that we take plywood and we put it under the nose wheel, which makes sense since the more you can disperse the wheel weight the less you will sink in. We were pulling with a [International Harvester] T-300 tug, and it had already started to get overstressed. It overheated a couple times and we had to decide whether we were we going to stop and have a chance of sinking more or were we going to keep going in. There was a whole group of people who didn't want to do anything except put down the plywood, but we had already pulled the T-300, so that wasn't a real option. There was some that said, "Why don't we just take the T-500, which is used to haul 747, and use it," and there was some debate and that went on for a while. Finally Deke gathered a bunch of people and asked us what we were going to do; everyone gave their input and what we agreed on was the T-500 hooked up to the T-300 to pull, and the plywood under the nose gear, and that's what we did. We learned at that point that you could no longer depend upon that metal ball hitting the ground and we had an Air Force captain who had spent some time doing that, studying lakebed and everything else.[12]

The second time we got stuck later on in the program we quickly decided we really need to do something, and that led to the extensive lakebed testing which matured into a vehicle with a rod that actually was sunk into the lakebed that measured displacement and could tell us actual hardness. That was done with JSC and the University of Kansas, later became Valenzuela Engineering that did that, and it helped the Air Force learn about lakebeds.

Gelzer: When did you shift away from your security role into a more management position?

Joe D: Prior to STS-1, I became chief of the management support branch. Up until then I had security, transportation and communications. At that point I took on photo lab, the mailroom, all of logistics, transportation, reproduction, and a new process that we had called word processing along with anything else that was assigned. So I had a big chunk and again a lot of us did it on the fly.

Gelzer: John [McKay] remembers being pulled off the F-104s and being sent down to work on shuttle every time that came up and then went back to his day job.

Joe D: That was the norm for everybody. John eventually did some control room work later on; the midnight shift. But again, we were in sort of a support role. People wanted us but they really didn't want us.

Gelzer: Because the people making the call here were JSC and KSC; they weren't Dryden?

Joe D: Yeah, and the rules we had. I had a tough time with them because we had started to go through the transition of [being under control of] Ames and the great time to look

[12]From early in the era of flight test programs at Edwards the Air Force relied on a standard 6lb steel ball lifted to 6 feet, then dropped onto the lakebed, to determine if the lakebed was dry enough to support aircraft operations. Investigators measured the diameter of the dimple left by the ball in the lakebed surface: larger than a certain size and the surface was still too soft for operations, smaller and it was ready for flight. Winter rains regularly flooded portions of the dry lake, sometimes even the entire lakebed, and since its more-than-44-square-mile surface represented the ideal landing area for an experimental aircraft, its accessibility was essential for unconventional operations.

at manpower, make reductions, look at more contractors. One of the things I said was: 'why can't we use some of the Dryden talent and JSC talent?' I say JSC because JSC pulled the 747; why couldn't they pull the orbiter. They had all the technical knowledge. You trusted them to take the 747 with the orbiter on top all over the place, literally move it within the facility, and yet on a landing we used our drivers, Dryden drivers, because we operated the T-300, but these drivers were always taking orders from someone else and it wasn't the upper management people, it was people from JSC that were just like them who probably had less experience. We towed airplanes around all the time, yet you had this great pooling of responsibilities and it was almost like they were saying: Dryden you're here to do what we want and that's all you're here for.

I didn't see so much of that from the administrative side of the house because a lot of us recognize we just couldn't do it with what we had. You just had to admit your limitations. A landing even then was very labor-intensive. You were on parking detail, you had to pull people away from flight ops that were in charge of avionics, key people, to run parking lots. You used the co-ops to do the little odds and ends but you had senior people doing a lot of things and doing it gladly, and yet some of them wondered whether management was really behind us. When Dryden actually made overtures for change or recognition of difficulties, Charlie Baker was one of those people who did to try to get more responsibility for Dryden; but we were actually rebuked.

Gelzer: – by Dryden management or by the other KSC, JSC?

Joe D: – by KSC specifically. JSC kind of agreed with us. I mean, few know that when you landed up at Vandenberg it was Dryden, Charlie Baker, and the contractors that went up and did the site surveys for the MSBLS and the PAPIs. When they were looking for work quotes, we felt we actually could do it cheaper than KSC, and when we were asked if we would bid on it, our bids were actually cheaper than the proposals by KSC. I had overall responsibility [and knew] the budget figures and I would say, "We could beat you every time," and they'd look at me and say, "You're crazy." I'd say, "No, I can beat you every time." "Well," they would argue, "but your contract cost are the same." My answer was, "You left out one important thing; we are already here. I can keep my TDY costs down and you can't. I mean, even to the point where you may not even be able to respond as quickly as we can." STS-3 proved that.

Gelzer: Talk about STS-3: they had an option for [landing] here or KSC and it was raining at KSC and water on the lakebed here, right?

Joe D: We had been concerned about water on the lakebed, but from very early on everyone felt comfortable and we would send pictures of the lakebed back to JSC showing the water dispersal. Always felt comfortable with landings on the lakebeds, the runways on the lakebed really dry out very quickly and most of it is not affected by water except runway 23. One of the principal water-gathering areas is the approach end of runway 15 and there's a stream that sort of runs in the far end of runway 23. So we were more concerned about the tow-back then the actual landing. There certainly was concern about the rain, if there was dampness on the lakebed what that would do. You might see some of this, mud so to speak -- they don't call it mud, surface material – pushed up into the orbiter because we always say that even on a nominal landing there is some degradation of tile caused by the lakebed. There's no getting around that – minor but it does occur.

So we had rain in the forecast, we actually did see a little rain prior to the scheduled landing and we had a meeting at roughly 4:30 -5 in the morning; a big weather briefing. Everyone was there to talk about landing. The discussion was: do we want to go to concrete, and what will we do if the concrete was wet and the lakebed was wet? Then we started getting into the degree of wetness, and for some reason people didn't feel comfortable with coming in for the landing on the concrete. The meeting was almost divided; there were people saying, "The forecast may not be as we think it is and things are looking better at White Sands," even though the front was supposed to be coming from that area. I never did understand that. That front was the front moving from the south pushing to the north, not quite northeast, refined. But right after that meeting the decision was made that we'd go to White Sands. It's important to note that one of the principal factors in the development of a White Sands plan consisted of two different options: an aircraft option, [Lockheed] C-5s, and a railroad option.[13]

Gelzer: You mean . . .?

Joe D: Carry the necessary equipment to that place. That leads to another key milestone, the OLF [Orbiter Lifting Frame]. That's important to note because you had developed processes and procedures to lift the orbiter, which are going to be important when the orbiters are delivered to some of these future resting places.[14]

The decision was made early in the morning for rail/ground operation. Dave Hoffman, referred to as Mr. KSC Railroad by some, was eager to put his rail plan into operation. Dave was the greatest lover of railroads you could imagine. I mean he was Mr. Railroad. He was responsible for planning the fabrication of two of the critical units that were used to transport the 534 and the 508 units, which are monsters.[15] You are looking at 133,000 pounds for one and the other one was almost 100,000. He had planned in his mind, he developed it and we conducted a test, physically, before STS-3. We had suggested using the Air Force ramp [at Edwards] because it was the only one in existence down by the flight line, and most of the planning was done by Dave, who worked with the Santa Fe railroad people in developing procedures, developing the plan for each piece of equipment that had to go.

During a trial run we moved the railroad car in to transfer the 534 units from the ramp to the railcar. Nowhere in the planning had anyone taken into consideration that a railroad car, when it connects to another railroad car, has a little ramp that folds down, but there's a difference of about three feet or more in some cases between the locking gear and the end of a railcar, to hook it up to the next one in that dock. We quickly looked at it and said, "We've got a one-inch piece of steel, solid steel that can do that." We got that steel piece down to that railhead and put it so it was the interface between the ramp and the railcar so that you could drive over it. We soon found out that we did not know how to transfer the big the unit onto that railcar. It overlapped each end of the car by about an inch and a half: tires were actually over the edge of the railcar. We

---

[13]The reference here is to the Mate-DeMate-Device (MDD). The one built at Dryden was designed so that it could be dismantled, shipped someplace, and reassembled for use, then dismantled again and returned to Dryden where it would once again be assembled. The designers contemplated moving the MDD from place to place either by a number of Lockheed C-5s or by rail, depending on where the device had to go. The MDD was, in fact, never disassembled after construction, and over time it was actually rendered immovable with additional fixtures and welding.

[14]For comparison, NASA estimated that "it would take about 19 C-17/C-5 aircraft sorties, a significant Navy sealift operation, and 450 NASA and contractor personnel to complete [a] turnaround" if an orbiter had to make a Transoceanic Abort Landing (TAL) and the MDD moved to support it. NASA Fact Sheet Space Shuttle Transoceanic Abort Landing Sites, [n.d.]

[15]These refer to purge and cooling units attached by umbilical hoses to the orbiter's back end after landing that provide cooling to the orbiter's electronics and experiments once the vehicle is powered down, and assure positive pressure is maintained in the cargo bay since the latter item is not hermetically sealed.

STS-3 diverted to Northrup Strip, White Sands Missile Range, because of rain at the primary and secondary landing sites. Lacking an MDD there, alternate arrangements had to be made. Cranes are lowering the slingback in order to lift *Columbia* in order that the SCA can be towed under it and the two mated for return to KSC. STS-3 post landing image courtesy of Joe D'Agostino

spent over four hours but could not get that 534 to line up. We tried backing it up and everything else.

With the decision to go to White Sands everyone geared up. We had people flying in from Florida but we weren't prepared, literally; Dave Hoffman wasn't here, and he was the author of this plan, the expert, so it fell upon Dryden. Dryden took the enormous task of getting all that stuff onto the railcars. I was in charge of that operation. I divided the team up into two parts, Art Nash, who I sent down to the back of the shipping and receiving at the Air Force area, where he handled all of the large steel items going into the gondola cars, and I was at the ramp railhead.

Gelzer: It's the one and only time that they'd ever shipped the equipment.

Joe D: They had forty railcars to do this with, in two waves. After we got the first few cars of stuff out to the railhead everyone departed -- I mean, they were gone. I got nervous like anyone would; we'd never put the shipment on and there was only me and a few USA [United Space Alliance] -- it might have been Rockwell at the time, probably Rockwell -- guys to do this. Everybody else was going home or getting ready to go home. They were running to get all their stuff to get ready to go and we were sitting out here with the responsibility to do all this, and the weather was turning bad.

Nobody was here from KSC. The first few cars were easy -- we had some difficulty

Now mated to the SCA, *Columbia* begins its return journey to the Cape from New Mexico. For years after the landing shuttle support crews at Dryden kept dealing with the fine gypsum that found its way into crevices of the equipment they shipped to the White Sands for the landing. STS-3 return photo courtesy of Joe D'Agostino

backing some of the equipment on, but overall, everything was going pretty normal and then it was time to load the two special cars, the 534 and the 508, and I was anxious. We did it in half an hour. We lined it up; I'm there in the middle of it directing him on, guiding the driver and we did it -- first time -- 30 minutes; got the other unit on, that was a snap, got them on, and now the weather started turning bad. It was the only day in the history of the program where I've got sun on my head and been rained on and even had some snow in the middle of the night. One of those that you just did everything you could -- you had to -- to get it ready to go.

Dave came about midnight, finally got here, and everything was almost all loaded. He paid me one of the greatest compliments I ever received. I looked at the operation and said, "Dave, it's yours; you're the boss, it's yours, here it is, I've done everything I can for you." He looked at me and said, "You've done such a great job, I need to get to White Sands," and left.

After less than four or five railcars I threw the plan out. It wasn't making sense. The reason? On paper it was the greatest thing in the world, but stuff wasn't arriving on schedule. I had all of this stuff lined up that was being shipped to me by the KSC people and I was supposed to put it on according to this railroad plan and the railcars didn't match [the planned loads]. You develop an eye, "Give me a 40-foot railcar," I'd say, line stuff up and I'd point to where I wanted this, this, that, and put it on. As a result, we had over 10 cars that, when we finished, were empty, and that wasn't according to the plan. But that enabled us to send a whole bunch of stuff that we weren't supposed to send, which proved to be a godsend. We sent a whole bunch of supplies, vehicles, and other equipment from Dryden that wasn't called for in the plan. We worked almost 40 hours straight getting the first cars ready and the train left on time. Everybody's gone at this point and we had the second group to wind up so Dave was at the other end. We got some rest finally after 40 hours.

Ironically, I got a call the next day saying, "Hey, your picture's in the paper. You made the front page of the Los Angeles *Times*." I did? Only problem is, the photographer was in back of me shooting me while I was guiding the traffic so you could only see my the bald spot on my head directing this thing.

That's an 800-mile trip that they had to make. After it was all over, getting the cars and everything else, the KSC director of transportation and logistics, Lex Pierce, came in and stated they needed to get to White Sands and so I arranged with my scheduling experience that I'd learned, to get them the KingAir to take them, and one of the nice gestures they did was ask me to come along.

Gelzer: Ames had no idea what you were doing?

Joe D: Right, they had no idea what we were doing whatsoever.

Gelzer: You're the one that told me that the day STS-1 was about to land here they called Ike [Gillam] and told him Dryden was no longer an independent center.

Joe D: Ike came in and told us he was no longer the director of Dryden, [Dr. Clarence "Sy"] Syverston was in charge of both organizations. That was a crushing blow.[16]

Gelzer: Ames had no idea what you guys were doing?

Joe D: No idea. To the point where, just to tell you how bad it was, the supervisor who was in charge of me, when he wrote my performance report, said there was some indication that I done a good job in shuttle stuff. I was one of three people that received honorable mention by NASA Headquarters and a photograph from the center director.

I went to White Sands. They wanted me out at the railhead because they had the same problem unloading it that Dryden had loading it. Theirs was worse: they actually had wooden telephone poles for their dock but it was elevated. I looked over their operation; I had nothing really physically to do

APOLLO 11 ASTRONAUT Mike Collins tried his hand at flying simulated space shuttle landings with Center pilot John Manke. Collins, who was command module pilot for the first lunar flight, is now head of the Smithsonian Museum and was visiting the Center as part of a tour of USAF Reserve general officers.

at White Sands. Gene gave me the photographer thing I told you about.

I did get the rare experience of having the payload to carry to the payload people. The payload consisted of the first horseflies that flew in space, so I got to see the dead ones and the ones flying, got to talk a little to the scientists about some of the things that they

---

[16]STS-1 landed April 14, 1981. The formal announcement regarding Dryden's demotion came on April 27 of that year.

had learned. I learned the problem was because the flies got so used to flying in space without having to flap their wings that they didn't flap their wings, so atrophy set in and they died.

We were staying at the Super 8 Hotel; right next door was the Holiday Inn. They had a nice little piano bar. After we retired things we would always go over and have a little drink; you have to remember I'm an admin now so I know all the admin people, the procurement people really well, the co-ops, I mean, that's your group. We're having a drink and this female shouts, "Joe! Joe! Is that you?" I'm looking around. Now, here I am in New Mexico, I don't know a female in New Mexico and someone's calling my name. Turns out it was one of our co-ops, a young lady who was going to a university there and she and her friends were at the next table, so we had them out to watch the landing. Turned out we didn't have the landing the first day. I kind of felt sorry because the wind was blowing so bad and it was just getting in your eyes and the landing was actually postponed.

So we did all those things: had the landing, our people did a lot in terms of helping and we developed a great rapport with White Sands folks, very close-knit group. Ironically, later on in the years it proved invaluable because White Sands and us were very similar, being treated as stepchildren, and we would cry on each other's shoulder. We came back with almost all the railcars, half of them, empty. That's how easy it was. I brought back stuff that we didn't have time to prepare to bring back. We picked up a couple barbecues that they had with steaks still in them. We never knew until we got back here; took these things to bring over and we had food in them.

Gelzer: Let's go back to STS-1 security. Apparently there were some people camped out underneath the approach path of the orbiter. There were of course the people that were trying to get across the lakebed as soon as it touch down

Joe D: For STS-1?

Gelzer: Right! But then later on they came across some people who were camped out underneath the approach path, and when they asked them about it they said, "Yeah, we were here before, nobody ever caught us."

Joe D: We had a couple incidents where we had the Air Force respond to people. One of the problems you have with the lakebed was that we had very little fencing. There was a sports car that got out [beyond the fence], there were a couple of them. Each one was fortunately caught in enough time. Harry [Talbot] can talk to you about the experience he had during one of the landings in which he and his group, who were responsible for security, came across a couple trailers that were inappropriately named; they should have said Chicken [Mustang] Ranch on them; people were enterprising. They were enterprising.

Videographer (Steve Parcel): What were they named?

Joe D: I don't know what they were named but they were the chicken ranches. The activity was definitely money-making.

Gelzer: A little entrepreneur there.

Joe D: There were several entrepreneurs. I was back at the ramp here and we were really busy. The weights and balances [hangar] had a lot of people sitting there. It went very well except for those few little incidents, but people responded. Other than that it was pretty uneventful.

Gelzer: Does one landing stick out anymore than any others for you, for any particular reason?

Joe D: Yes, but they're for different reasons. STS-2 saw the orbiter porpoising and that was a little bit much for me, having gone through the pilot-induced oscillation. It turned out that they were bleeding off energy and you got to understand that if they're coming in too hot they have to get rid of some energy, and they did that, but I didn't understand that at that time. Also, I knew that there was probably some concern from the medical people about the amount of fluid that the astronauts were taking; maybe they were getting into a condition where they weren't drinking enough water. I knew that because one of the other responsibilities I had was the monitor for the medical contract. Dr. Bill Winters, when he retired, decided the person who had the best experience was me, and decided I would be the contract monitor, so I had a plate full.

The brake problem we had in STS-7, in which we landed one hour after notification [was another]. A very small force did an excellent job. The team got that orbiter around with less than 40 people.

Gelzer: The original brakes had beryllium in them and they got rid of those eventually.

Joe D: Yes, they got bigger, stiffer axles, bigger brake pucks, and some other things. The next big event everyone talks about is STS-4, the presidential visit.

Roger Barnicki did an outstanding job in getting ready for that. One of which took us totally by surprise and then the last minute decision by [President Ronald] Reagan's advisor to bring all the people down from the hill to the viewing area because the concern was there weren't enough people to listen to the president give his speech.

That caused some nightmares, but the funniest part was when the FBI came in from Bakersfield. I gave him the briefing, he gave me his briefing, at the end of it he said: "I'm happy you've done everything you need to do." As he was leaving I said: "You forgot to ask me the most important question." He looked at me and I said: "You forgot to ask me if the guards are qualified, and how qualified they are with their weapons." You have to understand, our guard force was 20, maybe 30-some people with part-timers. Most of them were elderly, retired people and anyone would have questioned their marksmanship. You don't put that type of person with a weapon on a presidential detail. I sent the FBI over to the Air Force and we actually had Air Force people sitting up on top of the building with weapons and got a lot of NASA people concerned because they could see these guys walking up on the roof, but we needed it. We just had to have them.

Gelzer: Do you recall any, I won't call them disasters, but any events that took place that were less than happy or less than stellar? Bad memories?

Joe D: The first few flights, I think we were all ecstatic that we got the job done. Initially a lot of information wasn't shared with us by KSC. We haven't developed quite the rapport that we had with JSC flight directors. One of the first things I personally was involved with, a personal achievement I took pride in, was taking the first few flight directors that were brand-new on flights of Dryden/Edwards.

Gelzer: This was to show them the area, the base?

Joe D: The area and what we were doing. It almost got to the point where KSC didn't talk to JSC. It wasn't fashionable to be talking to flight directors because once the orbiter lands here it was JSC's job [to handle it] and so, "get out of my way."

Gelzer: As soon as they clear the towers it was JSC's airplane.

Joe D: Right, and we had worked closely with the flight directors – at least I did – from a planning point of view. I was responsible in those days for what we call program requirements documentation, which meant that all the items people had tasked us to do, I had to go to the people at Dryden and ask if they agreed or disagreed that we could support such and such. It ranged from providing sheet metal work, vehicles, and how many trailers we would provide them, to tables and chairs.

The only thing that drove me crazy was the public affairs people who always wanted to be right there, on the spot. When I became the lead of all the shuttle stuff, I made the comment that it is real difficult meeting the requirements of the security people and the public affairs people because they come from two different environments and the analogy I used was: the people in the operations are happy if the public affairs people are never near the runway, and the public affairs people aren't happy unless they're under the wheel of the orbiter during landing. Some of the decisions I made even later on, when I had the joint responsibility for both, didn't make either party happy because I had to find some kind of a medium and I had been used to the Air Force days where public affairs was part of the group and the team; at Dryden the security and the public affairs people seemed like, from day one, were on opposite ends. Later on it became a team effort.

Gelzer: John Booksha talked about having set up the original offices for the shuttle group, putting up a building inside a building. Essentially, he took an existing building and put an entirely new structure within it. That's how you remember it, too?

Joe D: Yes, AF building 4305, that's it. The requirements from KSC were so big that the Dryden facilities weren't able to accommodate everything. One of the solutions was to use North Base, because it had some storage facilities, a warehouse and a hangar that were available for immediate occupancy.

[Building] 4305 was great. They actually built all the buildings that currently still exist today if you go up there: a building within a building. The Rockwell people worked out at North Base. We provided a daily transportation from North Base to the MDD and shuttle hangar.

The shuttle hangar only consisted of the hangar, the warehouse, a small patch of land that got expanded three times, and trailers to include the holdovers from ALT including the tower that Pacific Telephone gave us for one dollar because they no longer wanted it. It served as the microwave link between the MDD, control room, and Palmdale.

Gelzer: In the time that you worked here on the shuttle program what has been your biggest challenge or obstacle?

Joe D: I think two: managing the budget for the shuttle, and contract renewals. I mean, when you're managing a contract that has to be renewed every three years, has to be rebid and awarded, it takes a lot of time and a lot of effort. Challenges that I looked at not as challenging included developing a rapport with the JSC people, and developing rapport with flight ops.

The greatest challenge from a personal point of view was trying to meet the requirements of people, KSC, JSC, and other customers, and trying to do a balancing act. It's difficult. If I invited you, a couple of other people to dinner, what do you serve? You might be allergic to certain things, I can't please you all, and when you have 500 to

1000 people it's tough, it really is, but you've got to get the job done because you only have one chance sometimes to do this.

So what do you do to the scientist that has spent 30 years or 40 years or 50 years trying to get an experiment in and you couldn't get him the nitrogen he needed to save his experiment because you needed that [elsewhere]? In one case one of the guys from New York brought all of his equipment here, 30 years of work, and forgot to bring the power interface cable. One silly little five-dollar cable would have ruined 30 years of work; we were able to get one from the avionics people to meet his requirements.

Gelzer: What did you have to do to make that possible?

Joe D: The Air Force had been using 23 on the lakebed so much, and we had so little rain, that it pounded the surface into sand. You see, water is the adhesive agent in clays. We knew all the requirements, knew what had to be made, so we moved the runway. I knew you could put a runway right next to the other one, so I did it. The only challenge was getting the people's buy-in. We knew the people from the geodetic survey here on the base, up at North Base; we knew all the players, it was just getting them together and going and doing it.

Gelzer: Taking one line off and putting another line down?

Joe D: And it's not like today. We have too many committees. Chief Willy Bell said it best. Chief Bell, on one of his visits after he retired, came by a meeting that we had with the Air Force to discuss what we were going to do on the convoy in preparation for a landing. The room was filled. There must have been 150 people in attendance. Bell walked by and I saw him and we walked out. He said: "Joe, I can't believe this; remember when we had that problem with the general and you and I got together and Deke [Slayton] said 'go do it' and we did it? Now they got a committee full of 150 people and none of them can make a decision." Times have changed; they really have. Your success has to be built on knowing what the rules and requirements are and still getting the work done.

On one of the landings I was asked to take a group of people out to visit the orbiter -- it was the KSC director that specifically asked for it. We got out there and you do your thing like any good person; you stand back and let the VIPs go ahead, after all, they are the people that run NASA, so to speak. I waited 'til all the proper times [but] the convoy commander, who works for the KSC director, doesn't even want his own boss to go out. I knew what was going on and said to him: "Bill, you got to let this guy out. Come on Bill, come on." I finally got through to him, and everything is settled -- they go out to walk to the orbiter and an astronaut walks out of the orbiter. I'm 300 yards back and the shuttle pilot walks directly by them and shakes my hand and says, "Thank you Joe." Years later I asked: "Why did you do that? It was almost embarrassing for me." He said: "I didn't know any of them. I knew you." So you get these little paybacks that are fun. It's like in any other job: it's the little things that don't seem like much, that are more important than the big, glorious things that you worked on.

Gordon Fullerton

Six years after entering Air Force C. Gordon Fullerton was selected to attend the Air Force Aerospace Research Pilot School (today the Air Force Test Pilot School), at Edwards Air Force Base, CA, in 1964. An aeronautical engineer by academic training, Fullerton was assigned to the Bomber Operations Division, at Wright-Patterson Air Force Base, Dayton, OH, where he flew every type of heavy aircraft in the U.S. inventory. He was chosen to be a crewmember on the Air Force's Manned Orbiting Laboratory program in 1966. Following its cancellation, Fullerton joined NASA's astronaut corps, in 1969. He served on the support crews for the Apollo 14, 15, 16 and 17 lunar missions. In 1977, Fullerton was assigned to one of the two flight crews that piloted the space shuttle prototype *Enterprise* during the Approach and Landing Test program at Dryden.

Early in the shuttle program Fullerton was one of the central motivators for and served as Instructor Pilot of the Heavy Aircraft Pilot Training program directed at would-be shuttle flight crews. This was because all the pilots selected as Apollo astronauts, and who were now residuals from that era, came from fighters; they had no experience in heavy aircraft. Because of this they had no experience working checklists with other crewmembers, an essential element of commanding and piloting the shuttle.

Fullerton was the pilot of *Columbia* on STS-3, March 22-30, 1982. That mission became the only space shuttle mission to land at White Sands, NM. He served as commander of STS-51F, on *Challenger*, in July 29, 1985. Following those missions, and eschewing a chance for an Air Force generalship, Fullerton opted to fly new and exciting things at Dryden. NASA's NB-52B had been parked for four years and one of his first jobs was to return the aircraft to flying status.

As the shuttle program advanced, problems surfaced. First among these that Fullerton worked on was the crosswind limit on tires for landing. The work involved one of NASA's Convair CV-990s.

When asked whether he felt he had been a pioneer in aviation he demurred, pointing to Fitz Fulton as an example of one. But then Fullerton admitted that the most pioneering thing he did was to fly the ALT, which was unique in every respect.

NASA Johnson Space Center Oral History Project Oral History Transcript, C. Gordon Fullerton, interviewed by Rebecca Wright, NASA Dryden Flight Research Center, California – 6 May 2002

Fullerton: The next long-term thing on the horizon was the shuttle. So, after Apollo 17 flew, I worked on the shuttle, and I worked cockpits and displays and controls. Always been of interest to me. Beside, I'd run across a lot of really crummy designs in learning to fly certain airplanes, and I thought I could do better. And so as it turned out, that was a real challenge to, with the shuttle, rather than lying on your back on the end of a rocket riding into space, you had possibility of controlling it, both in the vertical mode and coming back as an airplane pilot at the end. The whole complexity of it is far more complex than the rockets, as far as what the man could do.

So, putting all that together in a cockpit was really intriguing, and I enjoy working with stuff in an engineering sense, so it was perfect, and I became the cockpit design czar, sort of, to go to really organize and set up and go to all the reviews. I had a big foam core cardboard mockup of the entire cockpit built right there in the Astronaut Office, and I cycled all the other guys in there to say, "What can you see? What would

Enterprise in free-flight turning toward Rogers Dry Lake at Edwards Air Force Base. This is second of five free-flights, indicated by the presence of the tail cone on the orbiter. NASA ECN-8607

you do if this was a checklist? Can you reach it?" So I did a human factor study on all that.

What was great about the assignment was that as the shuttle was built, the first one, the *Enterprise*, I could see here's really what the drawings I signed off turned out to be.

Wright: Great feeling of accomplishment.

Fullerton: The other advantage was that to do that, you've got to know all the subsystems very well, to make any kind of intelligent decision about what the meter would say or the light should say, the nomenclature on the switches, all of that. That probably gave me a leg up on getting selected for the early *Enterprise* flights.

Wright: When did you learn, or how did you learn that you were going to be selected to be part of those test flights?

Fullerton: I should remember it as a stellar moment. It's probably George [W. S.] Abbey calling me up and saying, "Come over," and said, "How would you like to?" Dumb question; of course I'd like to. Generally that's how you found out about selections all through my time there. George Abbey was the head of the Flight Operations Directorate and the one who probably mostly decided and also told you.

Wright: And how did you train for these tests? What was your training before you started actually working with the *Enterprise*? Could you tell us about those experiences?

Fullerton: Well, if I'd never flown the *Enterprise*, doing the training was challenging and intriguing in its own right. People say, "How do you train?" thinking, well, you go to a school and somebody tells you how to do it. It's not that at all. Somebody's got to write the checklist, so you end up writing the checklist, working with each subsystems person and trying to come up with a pre-launch checklist for the approach and landing tests. So you're doing the work, that the learning comes from doing jobs that needed to be done. We worried about doing this dead-stick landing, so we had to train for that. I built a gadget to work on the T-38s that would allow you with any given weight to set the power with the speed brakes down to simulate what the data said the Orbiter would fly it at, so that we could go fly the pattern we intended to fly in T-38s, making steep descents, flaring, and touching down, and we did some of that right out here in the patterns that we flew the first tests.

The Shuttle Training Airplane, a Gulfstream II, was built as an airborne trainer, and so the four of us assigned to ALT served as the shuttle pilots along with a Gulfstream pilot to do many, many dives at the ground to get the STA, the Shuttle Training Aircraft, built and working right. And then the *Enterprise* was being built over here in Palmdale, and so Fred [W.] Haise [Jr.] and I flew many, many trips. I didn't even get out the chart to fly from Houston to El Paso, gas up, and go to Palmdale; I knew all the nav [navigation] aids and all the frequencies by heart. So we spent many hours in Palmdale in the *Enterprise* when they were running ground tests.

Wright: Did you feel that this was a role as an astronaut, or are you back to a role as a test pilot during these days?

Fullerton: The distinction is kind of blurred. Astronauts now, the orbiter's a pretty stable configuration, so they go to a school with ground school instructors that know the system, so they are astronauts in the—the pilots have got to learn the system, and the mission specialists have got to learn the payload and the flight plan. For ALT and then subsequently on the *Columbia*, we were clearly test pilots because we were doing stuff that there wasn't a procedure for. We were writing the procedure and then flying it for the first time.

Wright: And in this case, as you described it, you were partly a designer of helping to create those systems.

Fullerton: Yes, exactly. Very satisfying when you see really—I can go get in an orbiter right now, you know, and look at the panels, "Oh, yeah, I remember all this." It's a real feeling of personal pride, and the fact that it's still that way. They haven't changed it.

Wright: The simulator that, as you mentioned, the astronauts use to train with now, did you have an effect on how some of the simulations or some of the training equipment was set up for future astronauts as well?

Fullerton: Oh, yes. Since we were the first ones through the STA, you know, those procedures got developed, how we did it, based on us flying and trying. Still, you know, there have been changes over the years, but they're still doing that regularly. Now everybody, every crew that flies, flies a lot of STA flights in much the same way.

I thought it'd be really interesting—in fact, I kind of have set that up, too—and let's see, when I last flew it was 1985. So it's been sixteen years now, seventeen years since I last make an orbiter landing, and I'd like to just go get in the STA and grab the stick and try a landing, you know, simulating the interplanetary guy that's been on a sixteen-year voyage to Pluto or somewhere and comes back and has got to land it. I have this feeling I could do it. Of course, it's not like I've not flown a lot of airplanes since, but, you know, I'm going to try that sometime when they're out here.

Wright: Oh, good. I'd like to hear the results of that. That sounds really interesting. You were going through all these processes and procedures and training and creating all this as you were doing it, but at some point you learned of the day that actual tests were going to be held. Also, the amount of testing was cut, reduced back to only five of the tests when there was supposed to be many. Tell me about the special landing tests. Tell us about how that affected you, when you thought there might be more testing. Was that a good news that there was going to be less, or did you feel like that was a good decision?

Fullerton: What we were into at that time, the *Enterprise* and the approach and landing, the *Enterprise* was uniquely built to just do the approach and landing test. The idea, it would be sent back to the factory and all the space necessary, the systems would be put in it. That went by the board, never made that way.

But those initial tests, ALT was a program in itself, and there were a lot of people working on that, and money going into it that were holding up the *Columbia* coming along to do the first space flight. And so there was a constant debate about how many ALTs are enough, because this is holding up doing the real mission. And so the number of the possibilities—it turned out there were thirteen total flights. There were five captive, inert flights, they call it, where the orbiter was bolted on, completely inert, nothing moving, nothing running other than some instrumentation, and those flights, Fitz [Fitzhugh L.] Fulton [Jr.] and Tom [Thomas C.] McMurtry and flight engineers flew those five to the point where they said, "Okay, the combination is clear, and we understand what we've got here."

So then they decided to have some x number of captive, active flights, where the crew got on board and powered up the APUs [auxiliary power unit] and the electronics and all the subsystems, and those were dress rehearsals up to launch point. They had an open number of those. Turns out after three, they thought they'd learned all they needed were working. Had a couple of failures on number two, a big APU propellant leak. I was chasing that one.

Anyway, at three, they said, "Okay, it's time to go do it," and they were trying to get to the end as quick as possible, so they could get on with the *Columbia*. When we launched then, I flew on the first, third, and fifth of the tests. We did three with tail cone on, and Fred and I flew one and three, and then we took the tail cone off. It made a dramatic difference in the steepness of the glide slope. Joe [Joseph H.] Engle and Dick [Richard H.] Truly flew the first of those, landing out here on the lakebed.

And then the push was, "Let's have this—." It'd all gone quite well, although we discovered some serious design errors, but they were quickly fixed. So the grand finale then turned out to be free-flight five. Fred Haise and I landed on runway four going toward the lake out here, and we had a kind of an exciting landing there. It pointed up a flaw, really, in the design of the flight control software that led the pilot into a pilot-induced oscillation, and we bounced around and shocked a lot of people, probably more than—it didn't look that bad from inside the cockpit. But, again, that's why you do tests. You find out.

Then the debate was, should we fix that and test it some more. It was a strong feeling, like, that was a pretty exciting landing, which shouldn't be that exciting, or do we cut it off, fix it by testing and simulators, both airborne and on the ground. Do we know enough to press on? And it turned out that was the decision. You've got to cut the ALT off so we can go on the *Columbia* and get into orbit.

Wright: Could you share with us a few more details about what your roles were during those tests? I'm sure Fred Haise was the commander, and you were the pilot, for instance, landing or any of the other aspects. What exactly were you doing, and what were you having to be responsible for during those testing times?

Fullerton: Okay. The commander in the left seat primarily had the job of flying the airplane, to take the stick and fly it. There was a stick both places, so on each of the three flights, I got some of the flying time. But the prime role of the co-pilot was to take action when any of the subsystems had problems, monitor the systems. The pilot is busy watching where he's going and how he's doing on the profile, and checking the navigation displays and keeping the airplane on the profile we wanted to fly.

On the very first flight, the instant we pushed the button to blow the bolts and hop off the 747, the shock of that actually dislodged a little solder ball and a transistor on one of the computers, and we had the caution tone go off and the red light—I mean instantly. I'm looking, and we had three CRTs, [Cathode Ray Tubes] and one of those essentially went to halt, the one hooked to one of the four computers that monitored. This is pretty fundamental. All your control of the airplane is through fly-by wire and these computers.

So I had a cue card with a procedure if that happened, that we'd practiced in the simulator, and I had to turn around and pull some circuit breakers and throw a couple of switches to reduce your susceptibility to the next failure. I did that, and by the time I looked around, I realized, hey, this is flying pretty good, you know, because I was really distracted from the fundamental evaluation of the airplane at first.

That's roughly how the orbiter's set up. The guidance and control and fly on the airplane on a space reentry is designed by the cockpit and what displays are there, given to the left seat. The right seat's the co-pilot, and he's got access to the reaction control jets, the main engine, computers for space flight, for the auxiliary power units, the power, the hydraulics. All those critical supporting systems are over on the right side. Some are in the middle where both guys can grab. So all the landings you see, it's the commander's going to land it. He's not going to give that away, because you don't get very many.

Wright: How soon after the completion of the fifth test did you learn that you were going to become part of the STS-3 mission?

Fullerton: Oh, how soon was it? Now, the others were having a lot of trouble with the tile, the thermal protection system. They'd had fits and starts and failure of tests and delays. So it's a long time. The ALT was '77. The first launch was not till '81, right, four years later.

So what happened? During those four years, I picked some crews. The first crew that I was picked for was with Vance [D.] Brand. So I was his co-pilot, PLT, as we called it. I'm terrible for dates. I can't tell you just exactly how long it was. But then there was a reshuffle of things. No, that's not right. It was Fred Haise and I were on second flight, I think. Golly, I'd have to research this.

For a while I was going to fly with Fred. Then Fred decided he wasn't going to stick it out. He went off to management world with Grumman. So then I ended with Vance for a little while, and then finally with Jack [R.] Lousma, which was great. Jack's a great guy, and he'd flown on *Skylab*. He's not a test pilot, but very capable guy and a great guy to work with, and so I couldn't have done better to have a partner to fly with.

Wright: During that time period, were you training now in the simulators that you helped process originally?

Fullerton: Training, again, more engineering job than training job, because there were more details of the cockpit. The cockpit we had in ALT was just only the switches that applied. All the other systems now had to be put in. So I was back into that again. More reviews.

There were lots of changes, and then the software became a huge—the biggest stumbling block. The software that in these central computers not only control where you fly and control the flight path, but almost every other subsystem. And so getting the software wrung out and simulators writing the checklists, writing especially the malfunction procedures, what do you do if this breaks, if this breaks, if this light comes on. It's a book this thick of fine print, and amazingly, it's wrong most—you can get a room of the smartest people and you think we're going to get this right the first time, and then you go in the simulator and find out, whoops, that doesn't work, because it's a waterfall of interrelated effects every failure can be. And so we didn't really have it nailed down by STS-1.

There were lots of unknowns when STS-1 flew. There were lots of unknowns about the effect of a coolant loop failing and the cooling of the aft MDMs [multiplexer/demultiplexer], which was part of the data processing system. You know, a myriad of details. There were theories about what would happen, how the interaction would be, not really tested because there wasn't time. You just finally have to set a launch date and say, "We're going to go." You cannot be 100 percent sure of everything. And just bugs in the software. When we flew STS-3, we had another book this big called Program Notes, which were known flaws in the software. There was one subsystem that on the displays said "Off," because they'd gotten the polarity wrong and the logic, which they knew and they knew how to fix it, but we didn't fix it. We flew it that way, knowing that "Off" meant "On" for this subsystem. The crew had to train and keep all this in mind, because to fix it means you'd have to revalidate the whole software load again, and there wasn't time to do that. They had to call a halt and live with some real things you wouldn't live with if you'd bought a new car. That's all part of the challenge and excitement and satisfaction that comes with being involved with something brand new.

Wright: How was your confidence level in the orbiter and the whole process when you got ready to launch on STS-3? Did you feel it was ready to go?

Fullerton: Yes, but with this nagging thing, the thing that says "Off" when it's on, with a lot of cases where if this widget failed, this procedure in the malfunction book doesn't work in the simulator right. It doesn't come out right, and so you're flying knowing if this failed, there's going to be a lot of real-time conversation. There's not going to be a book answer, because it doesn't work in a simulator. It might be because the simulator's wrong. The simulator was a whole parallel development. We'd do an abort procedure and crash and burn, and we didn't know—well, is that because the simulator doesn't cope with this nor not? And the instructors didn't know because they were just as new at it as we were. And so, we, "Well, I hope it's a simulator problem." And so we'd write it and document it, and they'd take it off, and somebody would research it, and sometimes you'd get the answer, and sometimes you' just kind of go by the board because you're just too busy.

And so there's always an element in anything this complex, and that's the thing. It's really a complex vehicle. It really is. Even now I'm sure there's some question marks

Gordon Fullerton in the right seat of an Tupolev Tu-144L, the Soviet supersonic airliner. Fullerton was part of a group of NASA engineers and pilots that traveled to Russia to conduct a series of test flights using the Tu-144 in conjunction with the Russians. NASA EC98-44749-27

that exist there. When you're going to the nth detail about failures, if everything works like normal, it's all a piece of cake. It's when something breaks that you worry about, and is the big challenge to get to a point where you feel like you've got a handle on it.

So was I ready to not show up on the launch date? No, not at all. Was I quaking in my boots? No. Was I intense about the whole thing? Yes, mostly because I am worried about my part of this. Especially for pilots, it's the launch phase, because while it's short and concentrated, if anything goes wrong, the orbiter only takes care of the first failure. The second failure is pretty much left to the crew, generally, and so you worry about being ready to recognize a problem and do the right thing. You feel like the whole world's watching you when that failure occurs because of the manual action you've got to take to save the day. So it's that kind of pressure, pressure of performance, rather than fear or anything.

Wright: And you had spent a few minutes up in an orbiter, but yet you had never launched one. Would you like to share your experiences about the launch?

Fullerton: From *Enterprise* to *Columbia*? Yes. Well, the launch is a whole different ball game. I remember the first time, even though I'd spent a lot of time in the simulator, the simulators we built were fixed, one, and had the upstairs and the downstairs arranged horizontally, and then we had a two-seat, just the pilots' seats, in a motion base that would tip up and go up and down and shake around to simulate launch and entry. Those were the two orbiter trainer simulators. But most of the time they were both horizontal.

When I went to the Cape, I remember the first time when it's on the pad, and crawled in the hatch after being in my old cardboard, all these, and I was just flabbergasted how when you just rotated ninety degrees, how it becomes an entirely different outlook. I was lost. Wait a minute. Where's upstairs? Upstairs is this way. And so it's a huge psychological, physiological difference when you get on the pad and that whole part of it. You get over it, of course. You find yourself, "Wait a minute. I'm standing on an instrument panel. I'm not supposed to be standing on it." But that's the way it is. We knew we were going to do that. We built the switches recessed so you could stand on it. But that's a whole different thing.

Then, of course, the launch phase is like nothing, but your landing test is the last part of entry. So there was a familiarity there from ALT that certainly helped. But the eight days prior to entry was just a whole different world.

Wright: And while you were in orbit, one of the tasks that you had was to test the Remote Manipulator System [RMS]. Did you have a lot of training in that as well?

Fullerton: Yes, that was built by Spar Corporation, or whatever, Canadian firm. That was Canada's contribution, was the manipulator arm. So I went a couple of times up to Toronto to work with them on and to basically train, see how it worked. And then we had a full-size mockup at Houston with a 1G-capable arm driven by hydraulics. We had an electronic version of the arm, looking at screens in the windows and the simulator.

So there were a lot of tools to get the hang of working the arm. So that was pretty cool. I was prime on the STS-3. They had taken it out of the locks and waved it around a little on STS-2. Three, we actually grabbed something and picked it up and moved it around and put it back

Later, on [STS] 51-F, that same package we picked up and let go off of it and then went back and grabbed it. But Tony [Anthony W.] England did most of the arm work on 51-F.

Wright: Did you feel like the training and the actual tasks were close hand in hand?

Fullerton: Yes, we had good replication, so there were very few surprises. The nice thing about space flight, it's pretty pure. Airplanes fly through the air, and you've got air that does funny things and goes around corners differently, depending on the speed and all that. So simulations of airplane characteristics are much harder to do than when you're up there in a vacuum, where strictly Newton's laws are pretty pure up here and the predictions are very good.

Wright: While you were on that mission, you experienced a loss of appetite and some difficulty sleeping. Had you expected to have that kind of adjustment, or what were your expectations, being able to live in space?

Fullerton: On STS-3, that was, of course, my first look at it. STS-2, actually, they had some problems. They had a raw deal because they had a fuel cell—that was Engle and Truly—they had a fuel cell quit on them, and their planned five-day flight was axed to two and a half.

Of course, everybody has their acclimation problems. That's pretty consistent through the population. It takes about twenty-four hours to get to feel normal, at varying levels of discomfort. Most everybody can hang in there and do their stuff, even though they don't feel good. A few are pretty well debilitated. But they had not time, you know, in a two-and-a-half-day flight, they were cut short. By the time they got on orbit and traced down the problem and the decision was made to come back early, they were getting ready to come back. So they had no time other than to kind of respond, do things, that the ground was coming up, and they had some dizziness and orientation problems on entry that we learned about, and Jack and I worried about it a lot.

One thing that we did do, that I don't think they did, is we had a G-suit, like they wear in the F-18, except that for entry you could pump up the G-suit and just keep it that way, and so that helped you keep your blood flow up near your head, or assisted that. So we decided we're going to wear the G-suits. There was some controversy about whether you ought to pump them up or not, among individuals. We said, "We're going to pump them up."

The other thing about the motion sickness, we're not sure there's a direct correlation to flying airplanes and sickness. I know if you go up and do a lot of aerobatics day after day, you get to be much more tolerant of it. So Jack and I, we scheduled T-38 every chance we got in the last couple of weeks before we went down there, and I flew literally hundreds of aileron rolls. I know that's what would do it to me. If I did roll after roll after roll, I could make myself sick, and I did that, and I got to the point where it took hundreds of them to make me sick. But I did that figuring I don't know if this helps, but I had the opportunity, I'll do it, and the results were pretty much the same on both flights.

For the first day or so, I didn't ever throw up or anything. I never got disoriented, but I felt kind of fifty-fifty, you know. You're pretty happy to just—a malaise—you're happy to float around and relax rather than keep charging. And into

An official NASA photograph of Fullerton from the Approach and Landing Test program. He is holding the SCA with the obiter on top. Fullerton would fly *Enterprise, Columbia, Challenger,* and the Shuttle Carrier Aircraft once he retired from the astronaut corps and went to work at Dryden as a research pilot.

the second day, this is really fun and great, and you feel 100 percent. That was my—so whether the aileron rolls helped or not, I'm not sure, but it was relatively easy.

Wright: Where Engle and Truly's flight got cut, you had an extra day added on to yours because of the weather.

Fullerton: Right, so we had eight days, had seven scheduled and an extra one.

Wright: What were your thoughts when you heard mission control said—

Fullerton: "Wow!" We cheered. "Great!" Because we really had a busy time with just two people. This was an engineering test flight, and we had a flight plan full of stuff, and people fighting over, sticking in their stuff. So there was always something that you were watching the clock on. You had to do this coming up. We did have sleep periods, which we would use for window gazing, some part of it, because you don't need as much sleep as they were scheduling. But when they said, "Wave off," I remembered getting in the recycle book, going through the pages, shutting down some of the computers, opening the doors again, and I got all the way down, all of the sudden, I turned the page, and there was nothing on it, and there was this realization, hey, this is free time, and it was terrific.

We got out of the suits, and then we got something to eat and watched the world, and I wouldn't have had it any other way, if it had been my choice. In fact, we flew right over White Sands, where our landing site was. Just happened to be in the reentry attitude and we stayed in it. So we went half way around the world. The nose was pointing straight down, and as I looked up, I could see this monster dust storm going on there. It looked like it was all headed for Texas, the dust in the valley there. It was a clearly good decision. It looked really bad down there.

Wright: Yes, while you guys were having a, as you mentioned, free day, they were very busy down at White Sands preparing for your arrival.

Fullerton: Yes. Well, they were ready for us because we knew we were going there. This [Edwards AFB] was underwater out here. That's why they gave up on that.

Wright: Can you tell us about the landing? Was there anything different or any test procedures that you were working on with the landing that came in for STS-3? Anything different that you—

Fullerton: Well, where we planned to go, the main thing was this really fierce jet stream, fairly low altitude at 20,000 feet. The winds were over 100 knots out of the west, which is unusually high. John [W.] Young, I think, had flown some approaches in the STA ahead of time and decided if we made our normal left turn around to the southbound landing, coming from the west, we'd never make it back because of this wind blowing us away. So they changed to a single right turn, which put me on the inside of the turn, not Jack. It was clearly the right thing to do. So that was a wrinkle.

I could see the turn. He was asking me, "How's it look? How's it look?" because he was flying blind over there. I was saying, "Oh, it looks good. Keep it coming." So that was different. But we had lots of help figuring that out ahead of time.

The entry was pretty cool because it was an early morning landing, meaning that the main part of the reentry is at night, so we could see this glow from the ionization really bright out there. In fact, we had lost a couple of tiles on launch. We knew that because we'd looked out and had seen the holes in front of the windshield, and we looked at it with an arm camera. They said, "Not to worry. It's cool up on top there." We didn't know how many we'd lost from the bottom, but wasn't any use worrying about that. And then to see all this glow right there where the missing tiles were, gave us pause to think about it. Again, there was no point in worrying about it, nothing you can do. The spectacular light show through entry. Then the sun came up, which washes all that out, as it's dying out anyway. We went whistling by—and I spent four years at Davis Monthan [AFB] in Tucson—and as we did a roll reversal back to the right, I was looking down at Tucson going by and knew exactly what I was looking at. We were at about Mach 10. So it was a tour of the area of the country I knew. So, entry is really a great time for the pilots. You're flying. You're really flying. You're seeing where you're going. You're not just along for the ride at all.

Wright: And then you touched down without a problem.

Fullerton: Yes, the only problem there was a kind of a wheelie that Jack did. Again, it pointed out another flaw or room for improvement in the software. The gains between the stick and the elevons, that were good for flying up in the air, are away, were not good when the main wheels were on the ground, and he thought he had ballooned. He kind of planted it down but then came back on the stick, and the nose came up. So what? It didn't take off again, and we came down and rolled to a stop. A lot of people thought this is a terrible thing. I mean, we improved the software, and so people don't do that anymore, but we discovered a susceptibility. But other than that, we rolled to a stop, and we're out there surrounded by white gypsum.

   The family was there. It felt like I had been a long ways away. When I got down, we were on the ground, I'm feeling the gravity, it's all feeling normal, and I remember re-marking to Marie, my wife Marie, "You know, it was a terrific adventure. I'm here, but it feels like I've returned from somewhere a long way from here," you know, compared to flying in on an airplane. I guess it's true in a way, although you're going over all the time. But it's a great feeling, both space flights, too. I think it's a combination of—it's mostly a feeling not of relief that you're back. In a way, it's kind of crummy I'm down here slogging around in this gravity field, not nearly as much fun as floating. But the relief is that you got this huge team of people that are helping you through, and you're back, and it was a success, and you didn't screw up, do something to mess it up. That's a combination of good feelings, I remember, right out here on runway 23 on 51-F.

Gordon Fullerton (l) and Fred Haise prepare to speak with members of the media following their last flight in *Enterprise*, which sits in the background.

George Grimshaw

George Grimshaw was the final Space Shuttle Project and Ground Operations Manager at NASA Dryden. He began supporting the space shuttle program in 1979, two years before the launch of STS-1. By the end of the program he had supported 105 of the 135 space shuttle missions and shuttle related projects in a variety of technical, leadership and management roles. Grimshaw also supported the maintenance and modification of Dryden's aircraft fleet as an avionics manager and technician. Prior to working for NASA, he was a flight test electronics technician with three DoD aerospace contractors. Grimshaw also served as an instrument flight instructor on T-37 aircraft simulators for undergraduate pilots in the U.S. Air Force, and later as a part-time avionics technician with the California Air National Guard and U.S. Air Force Reserve.

George Grimshaw, interviewed by Erika Fedorko, Dryden Shuttle Documentary Interviews, July 2011

Fedorko: Let's start with you telling us your name and a little background about yourself.

Grimshaw: My name is George Grimshaw and I am currently the Dryden Shuttle Project and Operations Manager. I have been in aerospace for over 34 years. I started off my career in the Air Force, and afterward worked as a civilian at the Air Force Flight Test Center mission control center, where I first started supporting the shuttle program in 1979.[17] In 1980 we moved the mission control center to the current Ridley Mission Control Center, where I supported STS-1, -2, and -3. In 1984 I joined NASA at Dryden as an avionics technician and have been here ever since in varying roles, including supporting shuttle landings, recoveries and turnarounds.

Fedorko: What's involved in supporting shuttle landings at Dryden?

Grimshaw: We provide landing aid systems support, including the Microwave Scanning Beam-Landing System (MSBLS), which is the microwave frequency version of an instrument landing system that aircraft typically use for approach and landing. The MSBLS is used to align the orbiter with the runway as it begins its final approach. Once aligned with the runway, the Precision Approach Path Indicator (PAPI) lighting system provides the commander and pilot with a visual indication of the orbiter's position in relation to the threshold of the runway during final approach. As the orbiter nears the runway the ball-bar lighting system provides the commander and pilot with a final visual indication used to guide the orbiter to its touch down on the runway. Following landing we provide recovery and towing of the orbiter back to Dryden, which is then followed by shuttle post-landing processing using the Mate-Demate-Device (MDD) which is used to lift the orbiter for servicing and ultimately raises it so it can be mated on top of the 747 in preparation for ferry to Kennedy Space Center (KSC). The shuttle hangar and other facilities house the various vehicles and ground support equipment that are required for shuttle operations.

Fedorko: Could you take me through a day in your job? What kind of responsibilities do you have?

---

[17]Ridley Mission Control Center, named for Jack Linwood Ridley, pilot and project engineer on the Bell X-1.

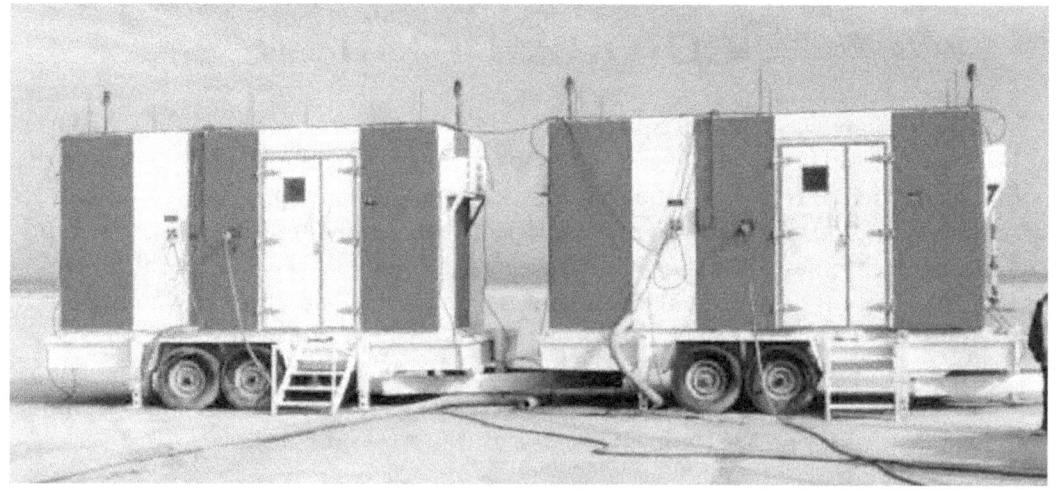

One of the Microwave Scanning Beam Landing Systems (MSBLS) located on the approach path to the main runway of the Rogers Dry Lake at Edwards AFB, CA.

Grimshaw: Basically, I am responsible for making sure that Dryden and Edwards AFB are prepared and ready to support orbiter landing, and post-flight support operations. This involves management and coordination with the launch and landing folks at KSC as well as the landing support folks at Johnson Space Center (JSC). Typically we coordinate any special requirements for an upcoming mission. We also look at landing times. Each day during a mission we have a Prime Landing Site (PLS) responsibility that's divided between Dryden, White Sands (Space Harbor, New Mexico) and Kennedy. About 70% of the time we are assigned PLS support. If something happens where the shuttle might need to come back and land, one or more of these sites will be available. Between missions we perform annual validations and preventative and unscheduled maintenance of the facilities, vehicles and ground support equipment to ensure Dryden's readiness to support.

Fedorko: Did you ever run in to any challenges with your work? Working with JSC, KSC? Is it difficult to coordinate?

Grimshaw: Yes. It can be difficult at times. Any time you are working with people no matter whether they're located at Dryden or across the country there are always challenges. There are different personalities and responsibilities. Sometimes the lines of responsibility blur: who is responsible for what. We continually work through these issues as they arise. Basically, we are a support site so we really look to the folks at KSC and JSC for guidance to make sure we understand their needs so we can better meet them.

Fedorko: Could you tell me about your early career when you were at the Air Force mission control preparing for STS-1?

Grimshaw: At the Air Force mission control center I worked in the ground station area where we received the radio data signals from aircraft undergoing flight tests. We processed those signals using computers and electronic equipment so that the data could be stripped out or reduced and then sent to control rooms to be displayed on strip charts. Flight test engineers could monitor and analyze the data being displayed, which included any of a number of parameters such as airspeed, altitude, flutter, engine performance, things of that nature. We provided this same support around the clock for STS-1.

Fedorko: Could you tell me about some of the people you worked with at Dryden? Does anyone stand out in your mind?

Grimshaw: From a shuttle standpoint the person who trained me for shuttle support was Charlie Baker. Charlie had the job that I currently have back at that time and Dryden had more responsibility then. We were the primary landing site. We practiced for landings a lot back in those days. I was a technician and driver for the convoy commander's vehicle which included leading the orbiter recovery convoy out to the various runway staging sites in preparation for a landing. This included all the lakebed runways as well as the main concrete runway 22 at Edwards Air Force Base. I spent quite a bit of time driving and learning the landmarks so that when we would position the convoy commander's vehicle, the rest of the convoy vehicles could position themselves based on where that vehicle was located in the staging area. This also involved learning safety zone distances, so once an orbiter landed we could move in to approximately 1,350 feet from it, and then once a safety assessment was completed and there were no toxic gases or other hazards present, we could move in to about 200 feet from the orbiter.

Fedorko: What types of toxic gases would you run in to?

Grimshaw: Mono-methyl-hydrazine is the first that comes to mind. We simply call it hydrazine or MMH. In training we were told that if you can smell it it's too late. It can have a really devastating effect on your health. Its primary purpose is to power the orbiter auxiliary power unit.

Fedorko: Why did Dryden switch from being a primary landing site to a backup landing site?

Grimshaw: The original intention was for Dryden to be a secondary landing site. Interesting that you would ask that question. A while back I was looking through some old inter-center agreements, between us and the other centers and I actually found the original document, dated September 30, 1974, for the design and construction of the Approach and Landing Test and secondary landing site facilities at Dryden. That really kind of tells the story as far as what the original purpose was for us -- being a secondary landing site.

For the first four missions, the research and development flight tests of the shuttle, we were designated as the primary landing site because of Rogers Dry Lake bed which provides a considerable safety margin should the orbiter need more room to land. So it kind of made sense from that stand-point to use Dryden.[18]

---

[18]The main runway at Edwards Air Force Base is roughly 15,000 feet long – long for any runway; yet this pales in comparison to the lakebed runways at Edwards, the longest of which is over 7 miles in length.

55

The shuttle recovery convoy heads down the more than a mile-long taxiway from Dryden to the main base past what is sometimes called "contractors' row." NASA EC05-0079-01

After we got a few missions under our belt the plan was to make Kennedy the primary landing site, which we did; but we were having some recurring tire wear and brake overheating problems so Dryden became the primary landing site once again. Once we got those issues resolved in the mid-'90's KSC became the primary landing site again and remained so.

Fedorko: Could you tell me about teamwork on the shuttle program? Was Dryden like a family?

Grimshaw: Yes, Dryden has always been like a family. You always hear about the Dryden family and I think that is one of the first things that impresses people when they hire on at Dryden.[19] I know it was that way with me. You also have closer-knit families based on project teams and certainly that was true with the shuttle. It was important that you not only knew what your job was but what other people's jobs were, what to look out for when you were out supporting a landing or recovery or in the turnaround / post-flight processing phase, just for safety purposes, just to make sure that if something didn't look right, if something wasn't going well, that people were looking out for people. So you do become a family when you work day in and day out with the folks that are supporting.

Fedorko: Did people use to joke around at all?

Grimshaw: Yes, there is always joking around and fun. We have an environment where people do things like that. It really indicates that you have a really strong healthy environment; that people are happy, they feel free to interact jokingly with others. So yes, there has been some joking that has gone on through the years.

Fedorko: Do you have any examples?

Grimshaw: I believe it was STS-61A, *Challenger's* last landing out at Edwards, that

[19]Dryden has always been one of the smaller NASA centers in terms of staff, hovering at about 1250 people total at the time of this writing. It was smaller still at the time Grimshaw speaks of.

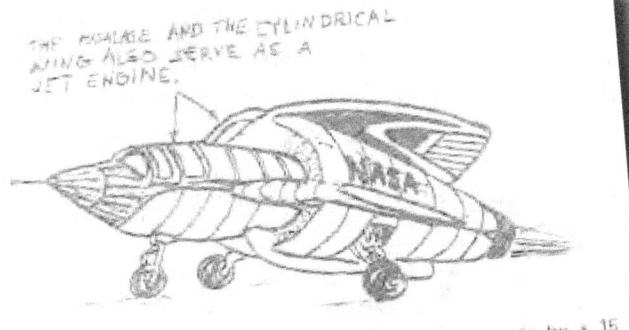

THE FUSALAGE AND THE CYLINDRICAL WING ALSO SERVE AS A JET ENGINE.

JET POWERED AIRCRAFT   This drawing was made by a 15 year old from Michigan. A more detailed drawing shows a heat resistant exhaust cone, compression chamber, hydraulic center, complete life support system and a test pilot.

would have been the fall of 1985. We landed on the lakebed and the decision was made that, rather than tow the orbiter back to Delta taxiway and get it on concrete as soon as possible to finish the tow back to the shuttle area, to take the shortcut and tow it on the lakebed around the compass rose and then on to the Dryden shuttle area ramp.[20] As we came cross the top of the compass rose the lakebed dirt was loose and the orbiter's tires began to sink into the sand. The orbiter had to eventually be stopped. We really couldn't tow it any further. So we're trying to figure out what to do and Charlie Baker ended up bringing out 1 inch thick sheets of plywood and we jacked up the main and the nose gear – the wheels on the orbiter – slid a piece of plywood underneath and then we put a piece of 4 foot by 8 foot plywood in front and probably for 100 yards we would roll it onto a piece of plywood, take the plywood that we had just rolled over, pull it out and take it in front. That is how we got the orbiter back to the ramp. Everybody had to pitch in to do that. It was really interesting to watch people come together and work. It didn't matter who you were or anything like that, and we were having fun with it. It was a serious time, but at the same time we were having fun with it because it was unique. It hadn't happened before. You could pretty much figure it wasn't going to happen again. You roll with the punches and do what you needed to do to make sure we got the orbiter back and got it back safely.

Fedorko: I thought Dryden had a way of measuring how stable the dirt was on the lakebed.

Grimshaw: Yeah, that's kind of part of the joke, because we did have a system that measured lakebed hardness and that test had been performed and evidently missed that particular area of the lakebed or got some faulty results from the test equipment.

Fedorko: Would you say that's your most memorable landing of the orbiter of do you recall any other ones that were particularly interesting?

Grimshaw: Having supported so many it's kind of hard to pick one. One that was very memorable was the Return To Flight from *Challenger*. STS-26 landed on lakebed runway 33, if I remember correctly, and Vice-President Bush came out. As we rolled the convoy across the lakebed to stage for landing we had to stop before crossing runway 22 to allow Air Force Two to land. After *Discovery* landed and we rolled up to the orbiter we immediately saw secret service agents at the command vehicle. One of them opened the door, came in and stood in front of the door inside of our vehicle as we waited for the crew to come out of the orbiter. Then a limousine with Vice-President Bush and Governor Deukmejian came up and they got out and greeted the crew and

[20]There is a compass rose, nearly a mile in diameter, at the edge of Rogers Dry Lake. The rose is located next to Dryden (research indicates the rose appeared sometime in the spring of 1953) with the cardinal direction West (270 degrees) ironically pointing directly at the shuttle area.

congratulated them on a successful mission. A stake truck with the media people on-board was driven in so they could take pictures of the vice-president, the crew and the governor. As soon as the vice-president and the governor were gone the secret service agent stepped out of the vehicle and we didn't see him anymore the rest of that recovery.

Fedorko: Do you recall meeting any other dignitaries or celebrities at shuttle landings?

Grimshaw: Other than meeting the crewmembers, I remember [NASA Administrator] Dan Goldin coming out after a landing when he was the administrator. It was a landing that was scheduled for Kennedy but ended up here at Dryden. He and his family happened to be at their home in the LA area and he drove to Dryden with his daughters to greet the crew and to see the orbiter. One of his daughters had her baby with her and he wanted to take both of his daughters out to show them the orbiter but the baby couldn't go with her for safety reasons. If you are going to be that close to the orbiter you have to be able to get away quickly. [Being able to escape at a moment's notice is a real issue since the orbiter still contains hazardous chemicals: though the safety crew has declared the vehicle "safe," sudden leaks or venting could occur without warning.] We couldn't take that risk so we ended up babysitting so his daughter could go out to the orbiter.

Fedorko: Could you tell me about the reactions to the *Columbia* and *Challenger* tragedies?

Grimshaw: *Challenger* is the one that is most cemented in my mind because we were working that morning. We had all the convoy elements in place and we were ready to go out and support a landing if need be. Our pre-operational checks were complete and we were standing by in a room where we could watch the launch on TV.

To provide a little bit more background, *Columbia* had just returned shortly before and we had been working a lot of overtime, so we had a lot of tired people going in to the STS-51-L mission. It was of one of those times when we were kicking back, closing our eyes a little bit while waiting for launch. As the shuttle launched and cleared the pad, we were cheering, and then a little over a minute after launch came the call "go at throttle up" then we saw something we had not seen before with these strange clouds of exhaust and we stared at the screen, not sure what we were seeing. We knew something was wrong and it got really quiet; nobody was saying a word. For me, I just kept waiting. I thought, "Well, obviously something has exploded, hopefully we will see an orbiter coming back around and landing at Kennedy." I don't know how many minutes we were there and we just kept waiting, and waiting, and waiting, and it never happened. Then the realization set in that it wasn't going to happen. Later, of course, we found out that, we had actually lost the orbiter and crew.

It was one of those events that you never forget. The rest of that day and probably for a few days it was like walking around in a fog. You had to come to work, you still had a job to do and you did it, but you did it with that cloud overhanging. It was

STA   This modified Gulfstream II is being flown by Approach and Landing Test pilots to familiarize themselves with the handling qualities of the Shuttle Orbiter prior to free flights this summer. The side action controls can be seen underneath the wings. These are similar to the side force generators on DFRC's Jetstar which were used in the passenger ride quality studies to put a sideward motion on the aircraft.

a very difficult time but I think that through the memorial service led by President Reagan in Houston, and the fact that it was a time of national mourning, it really helped. A lot of us were going through the same thing at the same time and it helped bring some closure.

Fedorko: What changed in the space shuttle program after that? How did it affect it?

Grimshaw: There was more emphasis on safety and listening to people when they noticed things that they didn't seem to think were quite right. I noticed this was already a part of the culture when I joined Dryden in 1984. If you see something that doesn't quite look right, say something. That was reinforced once again at Dryden.

The shuttle program had to go back and take a look at what caused the problem. They found out it was a solid rocket booster seal failure. That had to be fixed. Also there was more emphasis on management listening to the people who are the closest to the work being performed.

Training was also reemphasized. For example, there is a required course called "OV-220," which is required for anyone working in and around the orbiter. It is a four-hour familiarity course consisting of the danger points around the orbiter, including toxic gases and things of that nature. While it was designed as a one-time course, we had to re-take it prior to return to flight.

There was also a review of how we performed orbiter landing, recovery, and turn-around support at Dryden and making sure that we had the resources and processes we needed for safe and efficient operations when we had a landing at Dryden.

Fedorko: I've heard that people mention in the '80s Dryden was run quite a bit differently than KSC or JSC. Did you notice that as well?

Grimshaw: Yes. There was a big difference. Dryden was not as heavily tied to processes. We had the ability to go out and do what needed to do be done and come back and complete the paperwork later.

At Kennedy, especially with people working on the orbiter and related systems there, it was process and paperwork ad infinitum. To give you an example, when a technician performs a task on an orbiter they are provided a very detailed step-by-step work order. If a step could not be completed they had to stop work, contact the appropriate engineering authority, let them know what the problem with that step was and wait for further instructions and sometimes extensive reviews before continuing the assigned task.

At Dryden, our processes aren't that defined and we really rely on the expertise of our technicians and engineers to make many of these judgment calls. Not to say that it's looser here; it's just different. They are working on a multi-billion dollar spacecraft that flies beyond Earth's atmosphere and are more intricate. We work on aircraft that fly within the atmosphere and do not to need to have as stringent of requirements.

Fedorko: Would you say there is an increased efficiency at Dryden because of that style?

Grimshaw: Increased efficiency… yes, there is. We were talking about some of this the other day. People from KSC and JSC like to come out here and do a lot of their testing because it is much easier to get things done versus trying to do it back at Johnson or back at Kennedy.

Also, when you become a part of a culture you have a tendency to carry that culture's practices with you no matter where you go or what you do. If you're working around space shuttle orbiters all the time and you have to be that intricate about the details, then you have a tendency to want to do that when you come to a place like Dryden where we didn't have to have that level of detail. But overall, folks coming out here felt they had a greater degree of freedom to do things and get results quicker. There is a time and place for that and then of course there is a time and a place for being more stringent.

Fedorko: Were there any clashes with that difference in style, were they surprised when they came out to Dryden?

Grimshaw: At times. More so for people who had not been to Dryden before. You have to remember that the shuttle program has been around for 40 years and from an advocacy standpoint, back to the mid-to-late '60's. They were actually looking at a space-shuttle-type vehicle, what it would look like, what it would do, what capabilities that it would have.

In 1971, President Nixon authorized the space shuttle program and we had people, Milt Thompson, Ken Iliff, and other engineers and pilots that were involved in related projects here such as the X-15 and lifting bodies. I have been told that we had an influx of folks from Johnson and Kennedy that came to Dryden in the early '70's and we didn't have room for them on site so they had to build a facility for them inside a hangar at North Base. So when you have a history that goes back that far, when people are coming out in the '80's and '90's and now the 2000's, you've worked together before and you kind of understand what the differences are. I don't think there are a lot of surprises anymore. Sometimes you get into a little bit of a culture clash in doing things but you work through them, move on, and you get the results you need.

Fedorko: What's happening out at Dryden now that the shuttle program is ending?

Grimshaw: Part of our activity since 2007 has been getting ready for Shuttle Transition and Retirement. We have to disposition all of the shuttle assets. That includes all the facilities, vehicles, ground support equipment, and various systems. We have been working very hard on preparations with the program back at JSC as well as the folks at Kennedy. We are estimating that it is going to take about two years to disposition everything. It's kind of a bittersweet time because the ultimate plan is that probably

JSC Public Affairs Specialist Nicole Cloutier and George Grimshaw next to *Atlantis* following the landing of STS-117 at Dryden in June 2007.

George Grimshaw next to *Endeavour*, following the landing of STS-126 on runway 04 L at Edwards AFB. In the background, beneath the orbiter, is the NASA 25 Convoy Commander's Vehicle whose design and development Grimshaw led. Runway 04L/22R is a 12,000 x 200 foot asphalt runway, the shortest and narrowest runway ever used for landing an orbiter. *Endeavour* was the only orbiter to land on this runway. Photo courtesy of Nicole Cloudier

in about five years there may be no evidence that there ever was a shuttle facility at Dryden. All the stuff goes away, including the buildings.

The MDD is scheduled to be torn down sometime in 2012, since it won't be needed after the last shuttle mission ends and there are costs associated with keeping it up-to-date and ready to support. And, if you really don't need it anymore, at some point it has to go away. Thirty-five years ago the shuttle hangar, the Mate-Demate-Device and the shuttle logistics warehouse were under construction, and over the next five years they will be demolished and no longer exist. It's bittersweet. [Current plans are to retain the hangar and some of the other facilities for use by other programs at Dryden.]

Fedorko: How do you feel now that the space shuttle program is coming to an end?

Grimshaw: Once again it's bittersweet. The shuttle is the most amazing machine that has ever been invented and flown by man, and for the last 30 years it has just been amazing to watch it and look at all the things that have been accomplished, culminating with building a space station. Sometime after this mission [STS-135] it's really going to hit that that's it, there really is no more. It's a difficult time. On the other hand, having worked in aerospace for many years you see programs and aircraft come and go.

When I was first entering my career it was an exciting time with B-1 bombers and F-16s and F-15s and now you see those aircraft being phased out and F-22s and F-35s coming in. There is a life cycle and at times I have a tendency to focus on it and say it was great to be a part of that. I think we like to associate ourselves with something that is bigger than us and the space shuttle is going to be one of those things for thousands of people in America who can look back and proudly say "I was a part of that."

Fedorko: Do you feel that the program ended prematurely or it was the right time to move on?

Grimshaw: Yeah, I would have liked to see it continue on. However, given the fact that we built the International Space Station and that it is so expensive to operate, and it would be neat to see a woman or man on Mars, it's almost like it's time to move on. Let's go and let's do something else. Human space flight programs are very expensive and while it would have been neat to continue doing shuttle missions to the ISS and working towards the moon and Mars, at some point you need to cut the string and start moving and concentrating on the Mars mission rather than flying shuttles. I think that if we are going to do that then, yeah, the timing is right. If we're not [going to do these other things] then maybe we ought to keep the shuttle going a little longer.

Fedorko: Is there anything else you would like people to know?

Grimshaw: It is amazing how something like the space shuttle program touches different generations of people. For instance, for STS-1 both of my grandmothers were out here visiting. It was a really interesting for them. We all had watched the Apollo program. We had watched Neil Armstrong and the other astronauts walk on the moon. That was really cool, but it was really interesting to try to look at what was happening with the shuttle through their eyes. One of my grandmothers was a farm wife, my other grandmother was a homemaker and a seamstress. The shuttle was kind of out there for them, and then to have them so close by, to hear the double sonic boom, and to watch it on TV as it landed was really something.

My kids grew up with the shuttle. For them it's not as big a deal. They have been out to shuttle landings, have been a part of different events at Dryden, such as "Bring Your Children to Work day" and interestingly enough, neither one of my kids chose a career in aerospace. For them it is more of the same. It's interesting to look at how different generations of people view the shuttle program.

George Grimshaw egressing *Endeavour* following a crew module tour during STS-126's post-flight servicing. Photo by David McBride

John McKay

John McKay joined the U.S. Navy and served in Vietnam. Completing his service he took a part-time position with a contractor at Edwards Air Force Base helping set up equipment related to the approaching shuttle program. That job evolved into a full-time position, which became a career with NASA. McKay worked primarily as an Airframe and Powerplant mechanic on a host of different aircraft, from T-38s to the famed F-104s, and many more. Like many at Dryden, he was called on to do extra duty when the shuttle came to the center; unlike others, much of his tasking involved working with the orbiters and the astronauts.

John McKay, Interviewed by Christian Gelzer, Dryden Flight Research Center
Shuttle Documentary Interviews, October 19, 2010

Gelzer: I would like to start by you telling us who you are and some background about yourself?

McKay: My name is John McKay and I retired in 2005. I was a branch chief in aircraft maintenance. I came out here in October 1976 just a few months out of the Navy. I hired in for six months temporary. There was some money on the MSBLS Program [Microwave Scanning Beam Landing System] that they hired on a couple people and I was on for six months and then they extended me for six months. I was an aircraft mechanic. So that's how I got here and that's how I stayed.

Gelzer: What had you done in the Navy?

McKay: I was an aircraft mechanic. I was in the VF-24 and the fighter squadron, F-8s out of NAS Miramar and attached to the U.S.S Hancock. We spent two cruises over in Vietnam. I came back off of the last cruise and decommissioned both the squadron and the ship and flew the F-8s over to Davis Monthan and that's when the F-14s came out. I only had about six months left so I went over to North Island [CA] and spent my last six months over there in a VS squadron, sub-chasers.

Gelzer: The MSBLS?

McKay: The MSBLS is a landing device for the shuttle. It is almost like a TACAN [Tactical Air Navigation system] where it brings the orbiter in. I really don't know a lot about it but they had money for it. I don't know if they still use the MSBLS sites out there but I imagine they do. It's a landing aid for the shuttle. That's how they got some of their money together to hire people.

Gelzer: You hired in here and began working as an aircraft mechanic?

McKay: Right!

Gelzer: And that's how you finished your career, not as a mechanic but working in the aircraft operations?

McKay: I never moved out of Code O [Flight Operations].
Gelzer: Tell me about your experiences with the shuttle. Lets start, because you've

been out here even before the shuttle first flew and you would have witnessed the very beginning, with the landings, even the ALTs.

McKay: I got here when the T-38s out of Houston started showing up and the *Enterprise* hadn't been delivered yet. Myself, and a guy named Dean Bryant, who was a crew chief on our T-38s, the Dryden -38s, we would take care of those from Johnson when they came in. As the ALT Program started really spooling up we would get six to eight T-38s out here and we would take care of all of them. That was my involvement with the ALT Program. That was quite a program. There were a lot of people out here a lot of people doing different things.

Gelzer: Did you watch each landing?

McKay: Oh yeah!

Gelzer: Where were you during the 5th and final landing?

McKay: That one was for the runway. I was down by building 1600, outside.[21]

Gelzer: You could see the whole thing?

McKay: Oh yeah! We didn't miss anything. Those days you could move around a little bit; you didn't have quite the security you do today. I think by that time they had brought some Northrop guys in from Johnson to take care of some of the T-38s because it had become overwhelming. They had so many of them out here we didn't have enough people to take care of it.

Gelzer: 1600 was the big Air Force hangar, so you were down there to watch the approach and landing. Any memories of what people were saying around you after they watched *Enterprise* kind of bobble as it came in?

McKay: We were in a vehicle, there was a few of us mechanics, and we watched it. We knew something was wrong. That's all it was to me, it was a glider and an airplane. I

MSBLS TESTS...The Center's Jetstar has been taking part in preliminary verification tests of the Shuttle's Microwave Scanning Beam Landing System at Kennedy Space Center in Florida.

Jetstar Flies In MSBLS Tests

The Center's Jetstar will be returning tomorrow from the Kennedy Space Center in Florida where it was flown in system verification tests of the Microwave Scanning Beam Landing System (MSBLS) installed on the Shuttle runway at KSC.

System verification is primarily initial checkout of the airborne MSBLS/Jetstar system and the ground stations, along with laser systems, to verify that all work together. The processing system for the data received from these flights is also checked out.

According to project manager Bob Baron, it was planned to fly five test flights, with about 12 total hours of flight testing. Crew on the Jetstar for these flights was Don Mallick and Ray Young. If the systems check out, the Jetstar will be returning to KSC in June for commissioning tests.

The Jetstar performed similar checkout and commissioning tests on the MSBLS system installed on lakebed runway 17 here in 1976, prior to the Shuttle's Approach and Landing Test (ALT) program.

[21]Building [hangar] 1600 is a one of the largest aircraft hangars at Edwards, positioned on the taxiway connecting Dryden to the base's main runway.

really didn't realize at the time it was PIO. We just got back in the truck and I drove back down to Dryden. It was pretty interesting and we just kind of waited for the information to leak out.

. . .After seeing a science fiction movie about a lifting body pilot, Bruce Peterson, Research, now runs to work - at 56 mph - and is capable of pulling off Volkswagen doors. . . .

Gelzer: What's your next connection then to the shuttle, the shuttle program in general?

McKay: After the ALT Program there were several delays with the original STS-1 launch. What a lot of us did here were other duties as assigned. If they needed you to do something we would do it, and as the shuttle was spooling up we would still take care of the T-38s.

For STS-1 we all spent the night here, the day of the launch. I had a camper out here in front of the Integrated Support Facility (ISF). A little motor home I borrowed and a bunch of us just crashed out in there because we just didn't want to mess with the traffic during the day of landing. We stayed out there for three days. I think it was a three-day mission if I remember correctly. We stayed out here and we took care of the T-38s.

But it wasn't until after STS-1 that I got involved down at the shuttle area, and that's when Charlie Baker was brought back. Charlie Baker was an old crew chief on the X-15. Larry Barnett brought him back to kind of run the operations side of the shuttle down for the Dryden side. That consisted of working with and helping all of the Kennedy Space Center and the Johnson Space Center people that came out; we were a go-between with the Air Force. We didn't have a lot of contractors at the time. I was down there helping Charlie. That was my first initiation with that.

Gelzer: What were you doing?

McKay: At the very beginning I was doing a lot of stuff for the Air Force, dealing with things with the lakebed, helping the ground support guys that were dealing with the cooling and purge units and all that stuff. Just making sure all those kind of things got done. There were a lot of hands-on things because there was just so much to do. At that time nothing was set in stone; the program was just so young here. We hadn't dealt with too many space missions in the past.

I would go down and get the welder from the Air Force when they were having a problem with something out on the lakebed. I also started running what they called a lakebed tester. It was a unit that almost looked like an I-beam trailer that had the main wheels of the shuttle right in the middle. We would load that with 110lbs. of pig iron, which would simulate the weight of an orbiter on landing. I would take a T-300 tug, which you would tow a B-52 with, or maybe even the 747, and we would go up and down the lakebed just on the outside of the landing zone and check for the depth of depression and make sure the lakebed was good to accept the shuttle. We did that on runway 17, 23 and 15 on the lakebed.[22] It was things kind of like that.

Gelzer: It was just 110lbs or 110,000?

McKay: I'm sorry, it was 110,000 lbs. There was a lot of support stuff. After that, STS-2 came. By then we had started what they were going to call Abort Once Around Team (AOA) for Dryden. So we got a bunch of mechanics together, a couple avionics guys, if I remember correctly, and we were going to be trained to take care of the orbiter if they had an Abort Once Around; in case they couldn't get the payload bay doors open they

[22]An orbiter often weighed in excess of 200,000lbs; halving that weight and placing it on one set of wheels was enough to indicate the lakebed's suitability for a landing.

would have to come in. What they were looking at was, in the future, starting to save a little money even though STS-3 was next. They were looking at it being operational after four flights. They were going to be launching these guys left and right. Of course that didn't happen. But we were going to have a team here at Dryden that did various things: sniff the orbiter for toxic fumes, open up the hatch, tow the orbiter, do different things. We started assembling that team and I was the head of that team at the time. We didn't really get started with that until STS-4 because STS-3, of course didn't land here: it went to the White Sands Missile Range.

Gelzer: Who made the recovery of STS-3? Was it all KSC people who went to White Sands?

McKay: Oh no, there was a ton of Dryden people. Now, as far as the hands-on people there from Dryden, it was Charlie Baker, Joe D'Agostino and we took some other technicians.

Gelzer: Did you go?

McKay: No, I stayed back here and was kind of running the shuttle area down there. Whenever they needed something found out, or if something was left behind, I would make sure it got packed up and shipped out: you still never knew if it was going to land here. When it would land here everybody got going, I think half of Dryden was gone. Now, whether they needed that many people, I don't think so. But Las Cruces was full of Dryden people.

Gelzer: Once STS-4 was up and they realized they were going to continue to bring the orbiters here longer than they expected to, did you continue this dual role?

McKay: Yes we did. The T-38s started fading away as I spent more time down at the shuttle. Around this time I put in for a crew chief slot on one of the F-104's and I got that. The shuttle did have the priority. I would go down about two weeks before any kind of launch and we would start doing what they call pre-ops. The Dryden technicians for the Abort Once Around team were in charge of, and responsible for the hatch opening of the orbiter and towing the orbiter. There was a fan that we used that was actually brought out of an orchard up north, and we just strapped a couple fuel tanks to it and made it work.[23] We had a driver for NASA 25, which was the van that we used on the shuttle convoy.

Now those were just the civil servants. We had contractors here that were in charge of all the ground support equipment as well, and of course you had Charlie Baker and Joe D. that were in charge of the Dryden Shuttle Operations. There was a lot of people down there at the time and honestly, I'm not sure what all of them did. I was pretty low on the food chain at the time. I was just busy keeping up with what I had to take care of.

Gelzer: Did you enjoy your time as crew chief of the F-104s?

McKay:. Oh yes I did! It turned out when my dad passed away, he was a pilot - a test pilot at Dryden - they had a fly-over.[24] Tom McMurtry flew N812NA; that was the same

---

[23]This fan was mounted on a trailer and used to blow fumes away from the orbiter so that personnel could work around the shuttle preparing for the astronauts to exit the vehicle. The Orbital Maneuvering System pods and the cooling system on the orbiter could leak various toxic chemicals.

airplane I became a crew chief on. I was in the Navy when my dad passed away and when I came back here I became the crew chief for 812.[25] It was pretty special. I was just really humble to be part of Dryden. I still am today. It's just a great place. It has done so much for me. It's an exciting place to work.

Gelzer: Tell me more about the shuttle?

McKay: STS-4 went well. There were a lot of things that went on. We were building as we were learning and more people were getting involved and they found out that they were going to have to land here more often. That's when we started building up the area down there in the shuttle area. We only had a couple trailers to start with and it was just packed with people all the time. It was nothing like it is today. We just had a place to hang our hats and some desks and some tables and that pretty much what it was. Of course we had the hangar and there were offices in there.

We had some surprises. As the program went to STS 6 or 7, whichever one Sally Ride was on [STS-7], they were supposed to land back at the Cape but they didn't. They landed in the early morning right out here on runway 15, which is a spectacular thing to see, just standing on the ramp. Runway 15 is just right in front of us there.

Gelzer: It just passes right in front of Dryden, like a review as it comes in to touch down.

McKay: It does!

Gelzer: Runway 15 was the X-15 runway?

McKay: Correct. Of course, we were always out on the convoy so we always had a good shot of what was going on anyway.

There were hiccups that we had and this was the first time that they had a brake problem. That ended up being an historic thing. We were getting ready. I opened up the hatch that morning. In fact, I opened up most of them because I would rotate my guys so they would get a shot at doing that. That was always a real pleasant experience; it was just really smelly in there, I've got to tell you. You never knew what to expect. They were really happy to see you, really, really happy--except for the sick ones. They were happy to get out and stand up.

As we were getting ready to tow the orbiter back to the hangar I was in the White Room truck. That's what we use to open up the hatch with. It had an enclosure and you were able to seal off the area. I noticed there was some dirt building up in front of the right hand main gear of the orbiter: the wheels were rotating but you could tell they weren't rotating as fast as they should be because there was some drag going on there. So I called up NASA 25, to a Lockheed supervisor in there: I told him I thought there was a problem with the right hand main gear. He said: "no, everything looks fine." I have no idea how he would know that because he wasn't close. Right then the tow bar separated. The tow bar was designed and built in a way so that you didn't want to put too much tension and pull on the nose gear. The nose gear was just about at it max de-

[24]John B. "Jack" McKay, Sr. served as a research pilot at the NACA/NASA desert facility from 1951 to 1971. During his career he flew an extraordinary variety of aircraft, including the D-558-1 and -2, the X-1 series, and the X-15, in which he achieved astronaut status.

[25]N812NA was one of three Lockheed F-104s that NASA purchased new from the manufacturer, the aircraft never having gone to a military customer first, such as the U.S. Air Force, which was almost always the case for high performance aircraft.

sign as far as weight. That is one reason why I called up, but he said "no, everything's fine." When I said separate, it is a piston that comes out, a red flag goes up, and a horn goes off--you can hear it for miles around because they really want to make sure they get your attention. That's when we found out the brakes were dragging.

So we get out and everybody's looking at this situation and Charlie Baker, being an old aircraft mechanic, and me, being an aircraft mechanic, both of us said: "let's go ahead and crack the B-nut on that and allow the pressure to dissipate and we'll just roll it back to the barn." Wow! You would of thought that we'd just said the worst things in the world, because this is not an airplane--this is a space ship and you just don't do that!

So we waited out there until they brought in an engineer from Downey, [CA] I believe. They flew him in on a helicopter. The decision was: to remove the wheels out there on the lakebed. So we went to get plywood, ¾ inch plywood, which we had on standby anyway, to jack up that side and remove the main tires and the brakes and then reinstall the wheels to get it back to the shuttle area. I guess that was a good idea because then they had virgin brakes to look at and see exactly what happened. Now, this brake issue went on for several landings after that, such that those brakes had to be removed out on either the lakebed or the main runway.

Gelzer: The cause being over heating?

McKay: I don't know what they found out with the brakes.

Gelzer: This was before they put the drogue chute on?

McKay: Yes, that was one of the reasons why they did that. Of course then with our B-52 "Balls 8," we did the Drogue Chute Program for the shuttle, another Dryden kudo. The brake issue became a real problem because, in the early days when you had the secret DoD missions with the Air Force and other stuff, Edwards was very happy to accommodate whatever you had to do. But when it wasn't their mission they didn't want the runway locked up, and I can understand that. It became a real bone of contention for a while because it took awhile to jack the orbiter up on the axle jacks, get the tires off, and remove the brakes. It was about a four to five hour ordeal anyway.

Gelzer: And at the time Edwards had only one runway and they weren't using the lakebed for anything else they wanted.

McKay: Yeah, that was it. The lakebed was seldom used in the later years.

Gelzer: What else?

McKay: By this time we were getting ready for the night landings. We pulled every-thing out of the bag on that. We had an aircraft carrier ball bar set up on runway 22. It was something you would find at any naval base when they do Field Carrier Landing Practices (FCLPs), when they do simulated carrier landings and they'll fly against a ball bar. We'd brought that up. We were looking at different types of lights for how we were going to light the runway. We did that for quite awhile. I believe that it was STS-8 that Richard "Dick" Truly, who was the commander, landed at night at Edwards.

Gelzer: Could he see the ball bar or was that just something you guys toyed with and then dismissed?

McKay: We toyed with it then we dismissed it. I don't know why that happened. The

Enterprise has come to a stop on the lakebed and several support vehicles have gathered around it in preparation for towing it back to Dryden. Always an inert vehicle, there was never any need to hook up a purge unit to the orbiter. NASA photo

main thing were the xenon lights. Those were 8 million-candle power lights. We were doing a lot of flights on runway 17 out on the lakebed; this would be at 3 o'clock in the morning in the summer. They would bring the Shuttle Training Aircraft (STAs) out and fly against it. Bill Dana, one of our Dryden pilots, flew against them a lot in the F-104s and did the approaches against the different things that we would put out there.

We would have to have the xenon lights up at 25 ft. on runway 22 because it was shaped somewhat like a banana. In order to be able to get over that, you know it's 15,000 ft. long, at 7,500 ft. it starts moving back down a little bit. That was a lot of fun.

Gelzer: The xenon lights would actually light the entire runway?

McKay: Yes, but we had several of them and we had both on each side. We were doing this for STS 5 or 6, it was other things that we were doing in between these launches. I can't remember which one it was, maybe STS-7, that it was going to be an early landing.

I had the xenon lights--we only had two trucks of them, one on each side of the approach end of runway 22--and I was on the far side checking them out. I get down off the truck and we are probably an hour from landing--I believe they had already de-orbit

69

burned and that's when you just don't move--I get off the truck and I go walking around it and there are two kids sitting off of runway 22 that had traveled from Los Angeles and parked their car on a 140th and walked all night, and they were literally within 100 ft. of the runway laying on the ground! It was pretty cold that morning. I walked over to them and I said: "What the hell are you doing?" I've got to give it to them: these guys had what it takes to do that. They told me they were on STS-1 right under 23 in the lakebed: never got caught. No one ever saw them. They knew exactly where they were. You have to know where you were going; I don't know how they found out, but I went ahead and called NASA-1 and the Air Police came over and got them. I found out later they just scared the devil out of them. They let them go about four hours later.

Gelzer: George Grimshaw talks about a family driving across the lakebed

McKay: I forgot that one. That was good! That was STS-2. We were coming up on taxi-way D, we had just reached it and we stopped the convoy. I can't remember why they had done that because I wasn't really involved with STS-2 that much. We had stopped and we turned around, and there was this convertible with about four or five people in it. Evidently, they had entered the lakebed from the East shore somewhere. For STS-2 there were still over 300,000 people here to see the landing, and with all the AP's it was amazing! They drove right up. I don't know what happened to them.

Gelzer: George said they were taken away like your two guys and held for a while and had the you-know-what scared out of them and then they left.

McKay: They were just there to see the shuttle. That's what they said. They drove right up to it. I don't know how they got that close, because when you are doing shuttle-land-ing practices you've got a zillion helicopters out there, you got a lot of stuff going on. I was surprised that they were able to get that close.

Then I had another one. As the program progressed for us it wasn't an everyday situ-ation. We took it like it was the first landing every single time. By then the dignitaries stopped coming out and meeting the crew at the bottom of the stairs and that kind of stuff. This one time I had [Francis] "Dick" Scobee with me in the White Room. He was out with me since I was his ride in the convoy. We were waiting to get cleared to ap-proach the shuttle after it landed, and you could hear the APUs chugging once it came to a stop. They just chugged.

Gelzer: They are burping ammonia or?

McKay: Ammonia! Another time we were almost overcome by ammonia in the White Room. We got the fan on just in time.

Anyway, we were waiting to get cleared in, and I was going to drive the White Room up to the orbiter. I had my guy on top to line me up with the hatch and I just happened to look over. Of course, we were on the left-hand side of the orbiter, and I just look up and there was about a 15 ft. flame coming out of OMS pod. This was a steady burping flame that just looked like it was getting bigger, and they cleared me in right when I saw it. The same guy that told me there wasn't a problem with the main gear told me I was cleared to go in. Dick Scobee and I were looking at this, and I said: "I don't think I should be going in there: you have a fire in the left hand OMS pod coming out of the exhaust. That's hydrazine!" And he says: "No, everything looks good, you're cleared to go in." Dick looked at me and says: "Does it always happen like this?" "Oh, just a couple times." I radioed to the Lockheed guy in NASA 25: "You need stick your head out of the door and take a look at that." Evidently he did, and I don't know what they did after that, I think the shuttle crew just shut down. They have certain switches that shut down certain valves. But this time there was a broken hydrazine line in the OMS pod. [This happened at the end of STS-9/41A.][26]

Gelzer: That meant it was going to burn until the hydrazine was out?

McKay: Correct, or it got bigger. Never heard much more after that. We went up after they put it out. I don't know what would happen today if that happened.

Then we had another interesting little thing happen. Every one of the orbiters was different, they just were. They are hand built and they are unique and every door was unique.

When they made the tool to open up the hatch, the side hatch for the orbiter, like a lot of things in the space business, it was highly overly designed. This hatch tool cost $75,000 to have made and I believe there were three or four of them and those were all different. It was a T-handle and you had a button that you pushed. You put it in, it was a ½ inch drive socket, let's say that is what it resembled, and after you took off all the panels and the tiles that you had to take off to get the door secured, you have a handle that you must install, and then you have to plug in the mechanism which is to unlock the door. On the inside it is just like a hatch on a navy ship. They had these dogs that rolled over and that's how that was locked. You had to push in a button, install the T-handle and then you had to flip a tab over to unlock it. Now, the emergency way to open up that door would have been to take a ½ inch drive and put it in there and take a hammer and hit it. That would break the lock, and they would have to go in and fix it later, but you would be able to open up the hatch quickly.

I was up there again and my partner was the one doing the hatch handling this time and he was having a hard time with it. It reached the point that, when orbiter was coming in, I knew the serial number of the tool that worked the best on its hatch; hopefully KSC sent it out. I had given them this information about the peculiar hatches, and they told me that it wasn't true they're all the same. But from experience--and I have opened up the hatch a boatload of times--there was a difference. Greg, my partner that was trying to flip this latch over to unlock it and he was having a hard time, but he finally got it over. It just didn't seem right though. But you know, you just have to go with what you have, and as soon as he rotated the latch the whole handle came apart. There were springs and ball bearings that fell out onto the deck. That's something you would prefer not to call NASA 25 about, because now there is no way to get into the shuttle. If the crew was disabled in some way, then the only other way to get in there was to put a hole on the right hand side of the orbiter. There was an area there for which the Air Force fire department had this ram thing that they could shove in there and get inside. It would go through the orbiter itself. So Greg and I had the crew inside open it up from inside, which is not a big deal, although it would have been if they couldn't have done it, if they were incapacitated or something.

Gelzer: Were you in communication with the crew while you were in the White Room?

McKay: No, we never were. We just had radios to NASA 25. Most of the time there was a shield they would put over the window providing shade, so you're usually waving at them. So I call up NASA 25, and they get a hold of, however they do that--through, Johnson or whatever--and get the guys down in the mid deck to open up the hatch, which is kind of tricky. I believe they depressurize the orbiter, but it never

[26]Different forms of hydrazine are often used as monopropellants in spacecraft because they generate enormous energy from relatively small volumes of liquid. Regardless of the variant, all versions of hydrazine are extremely toxic.

seemed like that to me because, when you were rotating the dogs and unlatching the hatch, that's when the pressure would start relieving. The pressure would relieve right into your chest so you had to do it real slow because you felt like the door was being pushed. You had to be really careful how you opened that. You couldn't do it fast. You had to really take your time because the door weighed a tremendous amount.

Gelzer: What happened to the springs and the ball bearings?

McKay: They were on the White Room floor. We collected them as well as we could. That was a big deal, but evidently it wasn't big enough, because the only phone call I got was from an engineer back at Johnson. This guy had such an accent I could not understand half of what he said. After about 45 minutes trying to explain to him how it happened I finally said: "We are going to need to get together, or I'll just write it out." So what happened was, they redesigned it so that all you had to do was have a ½ inch drive. You push it in there and you open up the hatch. That was the fix—which they

The Space Shuttle *Columbia* on Rogers Dry Lake at Edwards AFB after landing to complete its first orbital mission on April 14, 1981. NASA EC 14786

could have done in the beginning.

Gelzer: Let's come back to the 2nd story, the window . . .

McKay: I can't remember the names. Was it George Nelson? It was the time we had the two un-tethered Manned Maneuvering Units fly [STS-13/41C, the 11th flight]. George Nelson had a nickname, Pinky, and Ox was the nickname of the other guy [James van Hoften], but I'm just thinking of Pinky. We got the White Room lined up, and for some reason it took us a long time to get up there, there were some issues and I wasn't cleared to go in. This time I was actually up in the White Room and I had my partner below driving the truck; we would have one of the Kennedy guys stand in front, right underneath the orbiter by the hatch, to help me line up. I would tell him and he would tell the driver, that's how we did it; it was really technical . . . Today we would have to have maybe 30 people to do that. I looked up, and I'm looking at the cockpit window and I see this butt pressed up against the glass, then he removes that, turns around and waves at me. I don't know how he did it, but he did—Pinky. It was great! You couldn't do that today. We got a big kick out of that. That was fun! [Ironically, "Dick" Scobee served on this mission . . .]

Gelzer: I heard tell of a landing on the lakebed that had water at some point and so they had to put planks down and move the orbiter and pull the planks up and put planks ahead of the tires and pull it.

McKay: Yeah, I think the first time that happened was STS-1 where the orbiter got stuck. This goes back to my 110,000lbs of pig iron rig. This is when people were held accountable for what they did and you took your job seriously, I believe everybody does today, but we didn't have thirty people riding on that tug with me; it was just me. I would report back if there was a problem, and there was this time, I was on the north side of lakebed runway 23 and I was using a lot of power to keep moving, and you could see from the indentation of those tires how far they were going down. Now, I never did measure it. I came off of the north side of 23 and I started heading back to the shuttle area--that's right across Compass Rose in front of Dryden. There, I sunk big time. I barely got that out of there. We were probably 4" deep in that area; they subsequently closed Compass Rose for a long time because of that. So I get back to the barn and I told Charlie Baker and Joe D. that we shouldn't be towing the orbiter through there; anywhere around there we are asking for problems, big problems. I think the program didn't like the idea of having to tow the orbiter all the way back to taxiway Delta, and then back down in front of all the hangars to Dryden. I think there were a lot of people that had a lot of different thoughts on that. To me, that was the safest way to do it. It was a tried and true journey and it if took another 20 minutes, half an hour, an hour - big deal.

    Johnson got the Air Force to do a hardness test, and I think they did a hardness test right out in front of the taxiway. They used this electronic stuff and this kind of goes back to my dad's twin.[27] He worked here as an engineer, and part of his work was the landing gear on the X-15, structural loads and that kind of stuff. What they would do back then is fly the Gooney Bird [C-47] to these different lakebeds that were used as emergency landing sites for the X-15. They would take a 6lb steel ball and drop it from 6', measure the indentation, and do the math. That's how they figured out if the lakebed was dry enough to support a landing. Well, the Air Force had all this electronic gear out there doing that, and I was told: "no, it's fine."

[27]Jim McKay, who was an aerospace engineeer and also worked at Dryden.

Well, we got it about third of the way back to the taxiway and the orbiter sank. We had to bring out the ¾ inch plywood. It was like building a pyramid. You had to get it up onto the plywood and just keep it rolling and had to keep taking the plywood from behind it and place it in front of the orbiter's wheels. That was pretty interesting. We got back and I said: "I told you so." They didn't want to hear it; that was too embarrassing.

There was another thing I was involved with that was interesting. We installed an inner glide-slope that consisted of a set of lights placed at 1,500' down the runway, on their left hand side of the runway. Or maybe it was 1,000' past the threshold, and then another 500' down farther was one light that stood up I think 15' in the air. I can't remember all the dimensions, but after they cleared the PAPI lights (Precision Approach Path Indicator) and were coming in and they leveled off, they were able to line the 1,000' with the 1,500' and they knew they were on the right glide-slope.

All of this stuff came in boxes from KSC. It was all in pieces, and they hired a couple people from the unemployment office to help me put all this together. We mounted these generators on carts. We went out to the bone yard and found a 12" stainless steel tube and did the math, and then we made fuel tanks out of those. I sat all of that up on the runway. Nobody, not one person, ever went out and checked the angle of the lights. We flew STA's against it when I got done one afternoon and we debriefed at the "office" on 90th and J [a restaurant/bar named Wing and a Prayer] over some beers, and they said it looked good, and that's what it was. Now there could have been somebody going out there but when I installed it I never had another person with me. It was just amazing.

Gelzer: Do you remember the SCA departure with the orbiter that had to come back because of an engine fire?

McKay: Vaguely, what year was that? In 1990 I acquired the HARV (High Angle of Attack Research Vehicle), F-18 as mechanic. That was a big project and we were just starting to install the thrust vectoring paddles on that, so my involvement down at the shuttle was starting to rollback. I still went down for about two or three launches in between that. I was on that bird from '90 to '96. Don't even think I was at work or around when they departed and had to come back. I know Pete Seidl was concerned. That's a national asset. They even had guards around the shuttle carrier aircraft for a long time.

Gelzer: They use to keep them apart?

McKay: Yes, for that reason. We had guards on it down here for a long time at night.

Gelzer: Guards inside of an Air Force base?

McKay: Yes, right. You never know who is going to belly crawl across that lakebed with a peashooter.

Gelzer: Ed Saltzman use to ride his bike across that.[28]

McKay: I know. I was just so fortunate to work on the shuttle program. It was almost like the old X-1 days or the old X-15 days; it was new, it hadn't been done, as reusable launch vehicle and landing. I had it down, from the de-orbit burn to landing was 59

[28]Edwin J. Saltzman came to the center as an engineer in 1954. He lived in North Edwards and rode to work on a bicycle, a commute of several miles each way, a ride made shorter if he cut across a portion of Rogers Dry Lake, something no longer possible.

minutes; I knew that when you heard it overhead it was three minutes before it would touchdown.

We were out on runway 17 one time, waiting for it, and the next thing you knew, right in the middle of the orbiter coming in, they changed their minds and went to runway 04. You should have seen all of us: that's not practiced, so we looked like a bunch of gazelles running across the Serengeti away from lions just trying to get over where the shuttle would be when it stopped. It was quite impressive.

Gelzer: And the convoy doesn't move fast.

McKay: I'm in the White Room and I'm not going to break it, so I'm just putting along. George Grimshaw and NASA 25, they're gone. They've got to get the big boys up there to see what's going on. I was just very fortunate to be a part of that. It was just a great thing. It really was. We were inventing things as we were going along and Charlie Baker, why, they couldn't have brought in a better guy. It was a lot of hard work.

Gelzer: Any major issues you recall, apart from the ones you've described?

McKay: I did have another one, which was interesting. I think it was Daniel "Dan" Brandenstein that had a landing later in the afternoon, and it was going to be on 22. I remember that he asked me to take a protractor and go out to the threshold of 22 and then tell him where the sun was at about at 4:31. So I went and got a regular protractor and held it up right on the horizon there, and told him exactly where the sun was. I called him back, told him, and he said: "thanks John." It was that kind of stuff.

Now, when the *Challenger* blew up, that was terrible. We were all in here at work waiting for the launch. We came in here for several mornings. As soon as we'd walk in the door they cancelled, so we'd turn around and go back home. It was to be a very, very early launch for us. I'll never forget when they finally went. We'd all sit together around these tables and there was a TV for us to watch. When it blew up Dan Bane turned around to me and said, "Is it suppose to do that?" I said: "No". Everything really hit the fan after that.

Gelzer: What happened here as a result or immediately after?

McKay: It wasn't like lock down.[29] We had our NASA guys here and whatever they were in charge of for recovery that morning, they collected the paperwork and put their vehicle or whatever they were doing to bed and went back to work here at their regular job.

Gelzer: You would have been working Return To Flight after *Challenger* because it was in the late '80s and you would have been there.

McKay: Yeah, we did. Return To Flight, that was interesting. The shuttle program decided everybody had to be re-educated on how things went, how things worked. We were going to have to have an official card to say we were checked out on the hatch, the towing, all of this, whereas before, we didn't have any of that. Dryden didn't go by those rules. In fact, we used our own stamps on the shuttle paperwork. We had desig-

---

[29]Typically, when an accident occurs during a significant flight at NASA, the control room doors are locked after the accident, preventing anyone from entering or leaving for a period of time. Investigators begin interviews to record events so nothing is lost because of time, and nothing brought into the control room is allowed out for a determined period.

End of an Era: Von Braun Dies

Dr. Wernher von Braun, former Deputy Associate Administrator of NASA and former Director of NASA's Marshall Space Flight Center, Huntsville, Alabama, is dead at 65. Death occured in a hospital in Alexandria, Virginia, following a lengthy illness.

Dr. von Braun's greatest contribution to the exploration of space has been the direction of the design and development of the Saturn family of launch vehicles. The Saturns sent men to the Moon in Project Apollo, placed the Skylab space station in Earth orbit, sent three separate crews to visit it for lengthy periods of experimentation, boosted American astronauts into Earth orbit for linkup with Russian cosmonauts in the Apollo-Soyuz Test Project, and orbited three Pegasus meteroid detection satellites.

Always interested in his adopted hometown of Huntsville, Dr. von Braun led other civic leaders in the establishment of the University of Alabama in the Huntsville Research Institute, the Research Park and the Alabama Space and Rocket Cen-

ter. Reflecting the esteem of his fellow townsmen, Huntsville's civic center was named the Von Braun Civic Center.

NASA's Acting Administrator Dr. Alan M. Lovelace made the following announcement, "We feel a deep sense of personal loss at the passing of Dr. von Braun, one of the world's outstanding pioneers in the field of space exploration. His integrity, his personal dedication to excellence, his personal faith in the future offer examples for all of us to emulate and pass on to future generations of Americans."

His name became synonymous with America in space when the team which he headed used a Jupiter-C rocket, developed in the U.S. Army missile program, to orbit Explorer 1, the western world's first Earth satellite on January 31, 1958.

nated inspectors here at Dryden so we used our own stamps on the shuttle paperwork. KSC had a fit about that but Charlie Baker backed all this up. He said: "Why give these guys stamps when they could be gone tomorrow. We might have some other people there. They have their stamps." .... It was decided that a bunch of us were going to go back to the Cape. We were going to get trained on brake removal, which we wouldn't of done anyway, the hatch opening, the cool and purge unit, which we never did even at the beginning (that was all KSC stuff). That's big stuff, that's all these guys did. We could have done that if we had spent more time on it. You have to remember that all these people at Dryden had other jobs here, so whenever the shuttle came in, work kind of stopped.

.... We went back to the Cape. Charlie Baker, myself, and a bunch of the guys, to get trained. It just so happened that that was when the *Discovery* left the Orbiter Processing Facility and went to the Vehicle Assembly Building. Unbeknownst to us, that was about the biggest deal in the world because that means it's leaving the garage and it's getting ready to go and fly. It was going to get stacked. Well, that was party time then. We could not get a hold of anybody to have the paperwork completed so we could get trained; everybody was out of contact, doing other stuff. So we went to the Cape for a whole week and had a vacation at Cocoa Beach. Now, we went to work everyday, but we didn't do much. We tried to do a few things but it just never worked.

Gelzer: Did you get certification?

McKay: No, we never did. Then, during the Return To Flight, the shuttle was going to come in here for that flight and they gave us a quality assurance person that had to witness the door opening and all that. That was all brand new. The young lady that they assigned to that task had never seen a hatch open before. There is not enough room in White Room anyway; there is too much stuff going on. You have a false floor that you have to put over the opening. You start with the floors up until you get the hatch down

Dr. Whitcomb Retires

The father of area rule, the Supercritical Wing, and Winglets that were all flight tested here has retired.

Dr. Richard Whitcomb, legendary NASA Aerodynamicist retired last month and will remain as a special consultant at NASA Langley for the next several months.

Area rule, or the coke bottle shape, allowed the early century series of fighters to fly supersonically and was conceived by Whitcomb in the early 1950s.

The Supercritical Wing was developed in the mid-1960s to reduce drag of transonic commercial airfoils and was tested here on a modified F-8.

Winglets, upward curving wingtips which save fuel, are currently being tested on a modified KC-135.

and you got all this stuff around. You've got Scott packs (self contained breathing apparatus) in there that you have to take care of and you have to be able to get to just in case there's a hydrazine or ammonia problem or some kind of thing going on to get these for you and crew that are inside.

So I told her: "Here is where you are going to stand, you're going to stand outside the door. You can look in the door if you want but that is where you are going to be and we are going to do our thing. When we get completely done, the crew comes out, they all go away, then you and I will sit down and we'll stamp the book off." Well, that was completely foreign to her. You don't stamp anything off after the fact: you do it you stamp it. You do it: you stamp it; you take these screws off, you stamp it. We never did that--you couldn't do that; it would take us two days to get the door open. The paperwork was just tremendous. She enjoyed that. She went along with it. She was a good sport.

Gelzer: Even though she never actually witnessed the door opening?

McKay: Yeah, that was kind of the thing. If they were really serious about this then maybe they should have given us somebody that was familiar with the hatch operation.

Gelzer: After you picked up on the HARV aircraft, did you still do anything with the landings?[30]

McKay: A couple, three times and then I kind of faded away. It was just a natural migration I think. We got too busy on the HARV, and then after that I had the SR-71, starting in '96 and I couldn't combine those two.

It was somewhat disappointing. By that time we had lost some slots here, guys had retired and there just weren't enough people to go around to support the Dryden side of that Abort Once Around. So that had gone away as well.

They still had the George Grimshaws' that were down there with the NASA 25 and they had to be there so if they had any problems with radios or anything they were right there.

Gelzer: We're just not staffed the way we once were. Is what you are saying?

---

[30]High Alpha Research Vehicle, an F-18 with thrust vectoring paddles at the exhaust nozzles, and nose strakes.

McKay: Correct! Now KSC takes care of that. They have people out here. Lets go back when Sally Ride landed here (STS-7 I believe): there was just a skeleton crew. By then the shuttle was operational.

Gelzer: It was not expected here?

McKay: It was not expected here, right. There were just a few people out here and boy, you should have seen how quick we got things done. Other than the brake situation from touchdown to the time that we were rolling, I would have to say was easily two hours. The whole enchilada, it was heaven. It's amazing what you can do with a skeleton crew.

Gelzer: And no supervision so to speak?

McKay: Less supervision.

Gelzer: You've seen pictures of the old days, the X-1 and X-15, and you've seen pictures of guys standing around an engine run on the X-1 and it's blowing flame out the back and they're just standing there with their fingers in their ears and their shirts off.

McKay: Back in those days you were allowed to use your imagination. You were seldom told "no." I believe that has changed a lot. I think the culture has changed. Everything is done by committee. The risk--people just don't want to go there and that's what space is, that's what flight research is--risk. You get a good team together and you know about each other and you learn about each other and you get to the point where you trust this team. You can do a lot of things. I just don't believe that that's happening today. There are a lot of things we couldn't get done today that we did. I got here in '76 so I missed out on a whole bunch of the early fun.

Gelzer: That's why you can still hear.

McKay: That's true! That's right!

Gelzer: When you were young did you go out to see your dad after he had landed the X-15?

McKay: I came out once. I believe I was nine, so that would make it '62. I believe that's how old I was, maybe 10. My older brother and I came out and I remember being up in the pilot's office and they aborted. There was a helicopter that wasn't in position, or it was something down range that they didn't have support on. He had gotten into the X-15 and he was down at the end of the runway, I think, or down at the end of the ramp on the way to a takeoff. I don't remember much of that. We never did get to come out again because of Mike Adams.[31] I remember coming out here, my mom and I. Mike and my dad went deer hunting and so my mom came out here at night to get my dad. That was about two weeks before that incident. They kind of put the kyboshes on it, or my dad did. I can't remember exactly. I never did get to see a real X-15 launch in the air. That's okay though.

Gelzer: Anything else you want to add?

[31] Air Force Maj. Mike Adams was killed on November 15, 1967 when his X-15 came apart in mid-air during reentry, the only fatality in the program.

McKay: I had a great time. I still have a great time. Dryden has done so much for me. It took a kid out of the navy. The people that were here at the time were just such great mentors. Everybody was into safety. It was just known that you were going to do the best that you possibly could every single day and that's how we looked at it.

Gelzer: You came out here about the same time the last lifting body flew?

McKay: I came right after it. In fact the lifting body was still down, I think, in one of the hangars.

Gelzer: The X-24B?

McKay: Yeah, they hadn't shipped it out yet and being so new I wasn't just going to walk down there. I had a great time and I'm still having a great time. I can't tell you all of it, the frickin parties. Killed a lot of brain cells.

Gelzer: Where did you guys go? The Wing and the Prayer?

McKay: No, do you know where The Britisher is? It's off of 10th and I. Betty Smith had a bar. It use to be a Vallarta, there's a bar in there. We still had the Le Basque on Lancaster Blvd. I don't know how we made it in to work a lot of the times but we did. Today I would never be able to do that.

Gelzer: Have you read Milt's book on the X-15?

McKay: Oh yeah!

Gelzer: It's starts early on with the infamous story of going out after work one night with your dad and an Air Force pilot. The spent all their cash and then somebody announced they had a Diner card so they were going to go to a place that accepted that and your dad went to sleep at the bar. They were still drinking 'til one in the morning. When they showed up to work the next day, I think it was Joe Vensel who wanted to know if everyone was capable to do their job, especially your dad; Milt said "oh, he went to bed at nine last night." Something to that effect so he covered his butt to make sure he was able to go and fly.

McKay: The one story that Milt didn't put in there was when he stayed out all night. He came home at about 4:30 or 5:00 in the morning and Terry woke up and said: "What are you doing?" He was on the edge of the bed and I think he was just undoing his shoes, and he said: "oh I'm just getting ready to go in sweetheart." Tied his shoe back up and he walked out. She had slept the whole night and had not a clue that he was not there. That's straight from his mouth. He came out and that's when we had a bed in the pilot's office over here.

Gelzer: He went to sleep there? There must not have been any flights that day.

McKay: I think he was going to stay home but that didn't quite work out. They had a good time. They definitely knew how to work hard and they played hard. That was the rule of the day. I think this place would be a little better for it if they lightened up some on that and still had flight parties. The first week I was here they had two flight parties. I think it was an F-111 flight party, and I can't remember the other one. Dave Scott, who was the center director, to the janitor was there and everybody was on a first name basis and everybody at about 8:30 or 9:00 would come around and everybody would get home and go to bed. That's when you got business done. That's when you were able to talk. Not that you couldn't talk out here then but it was just a looser atmosphere. Here I'm standing talking to a guy that rode the buggy on Apollo 15, having a beer with the guy that's cleaning the bathrooms. It was just a great thing! You wouldn't see that at a lot of other places but at Dryden you did. It was great!

Thomas C. McMurtry & Joseph H. Engle

Thomas C. McMurtry joined NASA in 1967 following extensive flight time with the U.S. Navy. Born in Crawfordsville, IN, in 1935. McMurtry received his BS in Mechanical Engineering from the University of Notre Dame in 1957 and is a graduate of the U.S. Navy Test Pilot School, Patuxent River, MD. At NASA, McMurtry worked as project pilot, flying on the AD-1 Oblique Wing program, the F-15 Digital Electronic Engine Control (DEEC) project, the KC-135 Winglets project and the F-8 Supercritical Wing program, among others. He also co-piloted the 747 Shuttle Carrier Aircraft as it transported the prototype shuttle *Enterprise* on its first launch in the Approach and Landing Tests, on August 12, 1977, and on ferry flights back to Florida. He rose to become Dryden's Chief Pilot and then Director for Flight Operations. McMurtry received the Iven C. Kincheloe Award from the Society of Experimental Test Pilots in recognition of his contributions as project pilot on the AD-1 Oblique Wing program. He retired from NASA's Dryden Flight Research Center after 32 years of service, accruing more than 11,000 hours of flying time in various aircraft including the U-2, the triple-sonic YF-12C, and the F-104.

Selected by Donald "Deke" Slayton as one of the four pilots to fly the prototype space shuttle Enterprise, Joe Engle offers the perspective from atop the 747 that McMurtry was flying.

Joe Engle completed pilot training with the U.S. Air Force in 1957. He graduated the U. S. Aerospace Research Pilot School in 1962 and was selected to fly the X-15 the following year; he stayed with that program until 1966, completing 16 flights, earning astronaut wings in that aircraft. He joined NASA's astronaut office in 1966 and served as a member of the backup crew on Apollo 14. With the cancellation of the Apollo program Engle became part of the cadre of astronauts responsible for shaping the shuttle. Chosen as one of the four pilots to fly the Approach and Landing Tests at Dryden using OV-101, the prototype shuttle *Enterprise*, Engle would command the shuttle *Columbia* on its second mission, STS-2, with Richard Truly. In August 1985 he commanded *Discovery* on STS-27, his last flight into space. Engle retired as a Major General in the USAF.

———

Tom McMurtry, interviewed by Guy T. Noffsinger, Shuttle Documentary Interviews, October 2010, edited April 2012.

Noffsinger: Your name and your title?

McMurtry: My name is Tom McMurtry and I was a research pilot, and the co-project pilot on the 747 shuttle Carrier Aircraft on the Approach and Landing Tests [ALT].

Noffsinger: Were you out here for the landing of the first orbiter?

McMurtry: I was fortunate to have been involved in the ALT, when we flew the shuttle *Enterprise* on top of the 747; there were five separation flights when we did that. [Richard] Dick Truly was one of the pilots involved, one of the astronauts involved in flying the shuttle *Enterprise* after it separated from the 747 and flew down and landed on the big lakebed. Joe Engle was also part of the ALT, but not John Young or Bob Crippen, who were the first crew to fly the shuttle into orbit and then come back and land on the lakebed. Because of the association of all the astronauts with the Approach and Landing Tests, I knew them. It was exciting to see the shuttle landing on the lakebed, knowing that they had actually been able to go into space and fly there, then come back and safely land at Edwards. It was exciting to have played a little part in that program.

IT'S WINGS DO WHAT?...Tom McMurty will be the project pilot on the AD-1 aircraft.

Noffsinger: What did you see in all these fans thousands and thousands of people and the National Guard and their helicopters. I mean, when that thing comes to a rolling stop and there is this huge collective sigh of relief and excitement.

McMurtry: It was a special day and anyone who was associated with the program or there just to see the shuttle return, felt a lot of pride in our country and our space program. Those emotions were finally released and you said: "Wow the flight was done safely and they're back home. The shuttle really does work. It's a great program. It's got a great future ahead of it."

Noffsinger: Did you work with Joe Engle when he was on the *Enterprise* and they were doing the separation tests?

McMurtry: Yes.

Noffsinger: I have his version of what was going on in the *Enterprise*: you were in the 747. Describe what you were thinking before, during and after the separation?

McMurtry: The first time the shuttle *Enterprise* separated from the 747 was the most exciting part of the program because, even though a lot of studies and analysis had been done and Fitzhugh "Fitz" Fulton and I, as the pilots in the 747, had done a lot of simulation work, there were still questions about whether or not it going to actually happen as we expected from those simulations and analysis. We had practiced the descents and we'd gone right up to the point where the shuttle would separate, so this was the final determination on how it was going to work. There was a very positive, loud thump when the shuttle separated. There was always the question of whether the shuttle was going to slide back and hit the tail of the 747 or not; we couldn't see tail but we could, after a few seconds, tell it didn't hit us and then the chase called "clear!" so we knew that the shuttle had separated high enough that it could then fly away from the 747. We could make a turn and come on back and land at Edwards. Honestly, that was probably the most exciting moment in the Approach and Landing Test, to feel and be there when the shuttle separated from the 747.

Noffsinger: If you could delve a little deeper, you are sitting there, you got your hands on the yoke and your hearing all this communication going through your ears, knowing what's about to happen, you've been training up to this point, you're just about to give it the big test: what are your thoughts? What are you hearing? If you could paint me a mental picture of the process of just before you guys make the decision that this is the real deal, to it actually happening. Did you look at it as it was flying away, or were your eyes buried in the instruments, where you looking at the horizon?

McMurtry: Because we had practiced and prepared for it, the intensity of it requires that you focus a 100% on what's going to happen, what you are doing and your part to play in it. That masked, I think, a lot of the emotions that some would expect one to have. In other words, I wasn't nervous, I wasn't thinking about anything other than: "I've got to do this and do what I was suppose to do right." So we went through the checklist, and Fitz pushed over and we accelerated. The shuttle crew actually separated the *Enterprise* from the 747 but we had to back them up; we could have actually separated them ourselves. Everything was tense in the airplane. I think that masked the emotions, but once you've done it and pulled away there was a big sigh of relief. I keep using that term: "Wow, we did it!" How exciting! We pulled away and the shuttle went the other way. We were able then to fly back. We flew over the shuttle as it landed on the lakebed and made a flyby. That was exciting too.

Noffsinger: Did you rock the wings a little bit?

McMurtry: I don't think we rocked the wings, but it was still exciting to see the 747 fly over the shuttle after it had successfully landed on the lakebed.

Noffsinger: Tell me about the day of the engine fire?

McMurtry: Gordon Fullerton and I were the two pilots in the airplane [747 SCA] at the time. Gordon was in the left seat and I was in the right seat. We had just taken off from Edwards and we were climbing up--the airplane didn't climb quickly, it was a fairly slow climber because it was very heavy. We had just gotten out over the Boron mine and, sure enough, our warning light came on the number three engine. We reduced the power on that engine and the light did not go out. We sent the engineer down to look at the engine and he could not observe anything at the time, but we felt that there was some problem with the engine so we went through the emergency procedures. We shut the engine down, declared an emergency, turned the airplane around and came back and landed the airplane. The fire warning light stayed on the whole time--until we were just ready to land, and just as we were touching down the light went out. So we all had a reason to breathe a sigh of relief as the airplane rolled out and came to a stop on the runway.

Noffsinger: What were your concerns and worries about, at the rate of speed and high airspeed it was coming in with, something that heavy on top of you, was there any concern about wobble? Can you take me back to what thoughts were in your head as you were holding on to that control as you were coming in turning the jet, dumping fuel, describe that for me.

McMurtry: Well, when a fire indication occurs in an airplane, that is about the worst thing that you can happen because things can burn off or burn further and cause a wing to actually separate from the airplane. Lots of bad things can hap-

Shot from a chase plane flying beneath it, the SCA carries *Enterprise* aloft. From this point of view the two aircraft almost seem one. NASA photo

pen with fires; that's just something you don't like to see happen. Even though it was associated with one particular engine, that engine could have actually caused a significant enough fire that the engine could have separated from the airplane; that would have created a much more serious handling problem with the airplane. As it was, the airplane is a good handling airplane, the engine didn't fall off, the fire was terminated, and we were able to turn around and land safely, and again we all sighed a breath of relief when we got on the runway.

Noffsinger: Was there a fire?

McMurtry: Yes, there was evidence afterwards that there was a fire, but it was not a significant fire. They took the engine apart, did a complete analysis and were unable to determine exactly caused the fire, but there was evidence that there was fire. So either some fuel leak or some oil leak--there was definitely a fire.

Noffsinger: Tell me your best memory over the entire program of your involvement in the space shuttle program.

McMurtry: It's hard to pin down one flight, the Approach and Landing Tests were very exciting because that was the start of the program. We had to demonstrate that the 747 could carry the shuttle. That was a major goal. But the other part was that we were actually going to separate from the shuttle and the shuttle was going to determine whether it was controllable and if it had the performance to fly down and land on the lakebed. Just being a part of that on an everyday basis and all the briefings that we attended, all the practices, the simulations that we participated in and then to fly the airplane.

I suppose I could pick out the very first flight that we flew with the 747 with the shuttle on top. There was no crew in *Enterprise* at that time. When we taxied out we'd made some taxi tests and Fitz felt that the airplane handled and performed adequately to go ahead and make the first take off. I never will forget that Fitz turned, and there were four people in the airplane, and shook hands with each of us, and I thought that was a pretty emotional thing to do. I get a little emotional when I think about that. Fitz was a great person to fly with. He was a wonderful pilot and also a very wonderful person. In fact he should be here telling stories because he's the one that really was the driver behind the 747 and the program. I was just fortunate to be a small part of it.

NASA Johnson space Center Oral History Project, Oral History Transcript, Joe H. Engle, Interviewed by Rebecca Wright, assisted by Sandra Johnson and Jennifer Ross-Nazzal, Houston, Texas 27 May 2004. (Excerpts).

Wright: 1977 began the taxi test for the shuttle, for the orbiter testing out at Dryden, and more plans were being made to test. Tell us how you became very involved in all of the testing procedures for the orbiter and how you were part of the Approach and Landing Test.

Engle: I had been selected as one of the two crews. [Richard H.] Dick Truly and I were one crew and Fred [W.] Haise [Jr.] and [C.] Gordon Fullerton were the other crew, who would fly the approach and landing tests, which were glide tests from off the top of [Boeing] 747. Initially, the concept was to put air-breathing engines on the orbiter and to take off and fly it around and check its aerodynamic characteristics and then return to land.

Also at one time, those engines were conceived as being able to be rotated inside the payload bay, or in part of the payload bay, so that after the reentry, they could be deployed and fired up and they could extend your landing area after reentering the atmosphere. That really wasn't a viable concept, it turned out, so the air-breathing engines never became part of the shuttle, although they were part of the Russian's Buran vehicle, which is a very close copy of our space shuttle. They flew it with air-breathing engines, jet engines, initially. But I think one of the reasons that I was selected to fly the shuttle, initially, by Deke was because of the experience that I'd had at Edwards with the X-15 and air launching from another vehicle, from a carrier vehicle, then glide testing unpowered glide testing of a low L/D, which means a low lift-to-drag-ratio airplane, not very much wing for a lot of drag, and that's what reentry vehicles, space vehicles, tend to do. They're not optimized aerodynamically, because they have to launch off a rocket and perform as a spacecraft in orbit, and a very small part of their operational mission is in the atmosphere acting like an airplane. So they're not optimized to glide shallow and land at very low speeds.

The initial Approach and Landing Tests on the orbiter vehicle were, in fact, just that, and that was to place the vehicle in aerodynamic flight by itself and exercise all of the systems that we could, hydraulic systems, electronic systems, the fight control systems, and landing gear, of course, systems and things like that, do it in a real flight environment, and to gather as much as flight test information as far as stability and control parameters and performance parameters, and do it partly in an ideal environment. In other words, not have to worry about coming in to land and the wind having coming up and giving you a big cross wind, or low clouds or things like that.

You could take off and an hour later, drop, and you knew what the weather was going to be. Of course, at Edwards, it was normally pretty good anyway. But you could set yourself up ideally over the lakebed, too, so you didn't have the navigation concerns that you do coming back from orbit. Plus, the vehicle itself, really, was ready to go before the rocket engines, the propulsion system was ready. So that gave NASA an opportunity to get a look at the orbiter vehicle, its basic configuration, its flight control system, and make sure that it had an airplane or vehicle that could fly the pattern, the approach, the flare, and the landing, which was a very, very small part of its mission, but a very, very critical part of its mission, and gain confidence in that prior to committing to launch into orbit.

So the Approach and Landing Test was designed to do just that. It was initially designed for about eleven flights, as I recall, and as we flew more and more flights, there was indeed pressure from the other end to hurry up and finish up so that those engineers could be assigned to the orbital flight test vehicle, and the orbital flight looked like it was going to take resources and shorten up the initial Approach and Landing Tests. The approach and landing tests were going very efficiently, too, so we were getting a lot of data and were able to condense the program from eleven flights down to what ended up to be five.

Wright: Were you involved in all the flights in some way or another? Were you either observing or chase planes?

Engle: Yes. Dick and I, when we were not flying, we normally would be in Mission Control here at Houston. The chase planes were flown by a cadre of pilots from the Astronaut Office, who were given the task of chase pilots, and they concentrated and practiced on chase techniques, joining up, flying formation, optimizing their relative position to the orbiter, particularly being able to adjust and maneuver in case there were certain things that needed to be observed and verified or confirmed to the crew or to ground, whether either a control surface in the right position or that nothing had fallen off the airplane. But then as the flare and the float into landing and a touchdown approached, they would verify that the landing gear in fact was coming down and was down and locked, then at that time would call out the height of the main gear above the ground as we came in to touch down.

So, developing proficiency in those areas so that they could do that kind of second nature, kind of intuitively or instinctively, and concentrate more on what they were looking at and not flying their chase airplane, keeping it in position, was a real important thing.

At that time, I don't think we had the radar altimeter working initially, in fact. But even when it was hooked up, we really didn't have confidence in it yet. So a visual callout on how many feet off the ground the main gear was, was very important. Because in the shuttle, you were so high off the ground, with not a lot of visibility out those windows, you had no feel for how far off the ground your wheels were, really; you just didn't know. So it was a big help.

The Approach and Landing Test crews pose for a picture in the nozzle of one of the Space Shuttle Main Engine (SSME) mockups at Dryden. L-R: Gordon Fullerton, Fred Haise, Richard Truly, and Joe Engle. NASA photo

Wright: While you were in the Mission Control with Dick Truly, what were some of the duties that you had? What were you looking for?

Engle: Primarily systems failures or anomalies that would come up. With things that would happen, whether or not it was okay to press on with the drop, or to continue the climb out to the drop, or what the options should be when that happened.

Quite frankly, with the tail cone on, the streamlined tail cone on the back end, that was there really to cut down on drag for ferry missions, and it is, in fact, employed on all missions when the bird is ferried either west coast back to east for launch, or east coast back to west for the major modifications, or major maintenance and mods. The orbiter flew pretty benignly with that tail cone on. You had a relatively shallow glide slope. You could get to a higher altitude for launch, because there was less drag, and the flight duration, as I recall, was well over five minutes. In fact, I think it was maybe up to seven minutes with the tail cone on.

But that was not the configuration that we needed to really have confidence in, in order to commit for an orbital launch, because that reentry and landing would be made with the engines exposed and the blunt tail and required a much steeper glide slope, much more demanding profile, much more condensed time period from flare to touchdown, because the air speed would bleed off much faster with the additional drag.

So although we were able to get a lot of really good systems data and time on the systems, on the hydraulic systems and electrical systems and computers and flight control system, although we were getting more time on those systems with the tail cone on, from a performance standpoint and a piloting task standpoint, we really didn't have what we needed until we flew it tail cone off, and those flights were only about two and a half minutes long.

Wright: Being a former X-15 pilot, you had been used to being dropped, whereas now the shuttle was actually being launched. Share with us the differences and what that experience was like to be in that type of maneuver.

Engle: Oh, it wasn't a lot different, really. It's a matter of going from a mated or a purely dependent situation on the carrier pilot to either throwing a switch or, in the case of the shuttle, pushing a button and blowing bolts and being free on your own. The X-15, your immediate concerns were, first of all, to keep it under control as you came off the hooks, but then to get the rocket engine lit right away so that you could get off on your mission, but you were dropping away underneath and there was no real concern with recontacting the carrier airplane.

With the space shuttle, that was our main concern, was to develop a separation maneuver with [Fitzhugh L.] "Fitz" Fulton [Jr.], the carrier pilot, to optimize the separation between the two vehicles, both vertically and laterally. So Fitz would put the combination vehicles in a slight dive to get the right air speed—you wanted to be able to get 240 knots, I believe it was—in level flight, with the tail cone off.

So he would dive the airplane and when he got on speed, he would call, — On speed. We would separate and call the separation, the — three, two, one, sep. And at that time, the orbiter was sitting with a 15-degree angle of incidence. In other words, it had 15 degrees angle of attack and was trying to fly off the 747 at that time. Fitz would, in addition, put the spoilers out to dump lift on the 747, throttle back to idle so that as we came off, we didn't slide back and take his tail off. Then the two of us would turn in different directions as well, so that as soon as we lost energy and started to come back down, we didn't come back down on top of him.

In looking at the videos, there was lots of room, lots of separation, but initially we weren't sure, so we optimized everything we could. We didn't compromise anything by doing that, but we did have plenty of room for separation, and it was a coordinated maneuver. Fitz would dive away to the side; we would pull off to the right; he'd dive off to the left, then we'd go wings level and go right into the data-gathering maneuvers, because we had very little time to get data. We had about a minute and a half to get data and then, well, the rest of the time was flare and land, to get the gear down and touch down.

Wright: Did you land on both the dry lakebed as well as on the concrete?

Engle: I landed only on the lakebed. The last flight, the second flight, tail cone off, Fred Haise and Gordo [Gordon Fullerton], the flight was set up to land on the runway, and I think one of the reasons was we had been able to gather enough, certainly the majority of the flight test data parameters that we needed for confidence that the wind-tunnel data was good and the analysis was good on the airplane, so we were able to get a lot of data on Free Flight 4.

Free Flight 5 was dedicated and designed as a demonstration maneuver that we can land the orbiter on a runway, and that was really the only purpose for Free Flight 5. So Fred and Gordo were set up to land on the runway.

Wright: How much did you and Dick Truly and Haise and Fullerton exchange information? How were you able to share the lessons learned from each of the crews with the other crew?

Engle: We would debrief, get together for sessions in the office and, if necessary, go to the simulator to show or demonstrate things that had happened or we thought had happened on the flight so there were no surprises for the next crew. There weren't a lot of flights, really, to pass that information on. A lot of times, the first time you see something in a flight, you're not sure what it was you saw or just how to describe it. We didn't have the luxury of that, really, because Fred and Gordo had a flight, tail cone on; Dick and I got the next one, and then Fred and Gordo flew the third one.

The third flight, really, was to engage the autopilot and to give the autopilot as much time to control and watch its reaction and correction to offsets on the final glide slope, and that really had no application to what Dick and I were going to do on the subsequent flight, the tail-cone-off flight, where we came down very, very steep and got the data with the tail cone off.

So, basic characteristics as far as handling, response of the vehicle, we were getting a pretty good feel for that through the simulators, both the ground-based simulators and the in- flight simulator, the Gulfstream G-2. So we did exchange what information was important. But each mission was a little unique during ALT [Approach and Landing Test].

Looking less impish in a more formal pose, the SCA crew sits around the SCA/orbiter model. From l to r: Tom McMurtry, Fitzhugh "Fitz" Fulton, and Victor "Vic" Horton. NASA E76-29620

John McTigue

John McTigue came to the NACA High Speed Flight Research Station in 1952. Among the projects he first worked on was the Douglas D-558-2 *Skystreak*, the first aircraft to reach Mach 2. McTigue moved to the X-15 program and soon became lead engineer on one of the three aircraft. Even while that program flew to new heights and speeds he moved to the lifting body program, one of the most unique flying vehicles ever conceived. He oversaw the engineering team on the heavy lifting bodies program. He was eventually made Program Manager and assigned to the Approach and Landing Tests.

Part 1 of the interview with John McTigue was conducted by Guy Noffsinger, Shuttle Documentary Interviews, October 2010.

Part 2 of the interview with John McTigue was conducted by Curtis Peebles, Dryden Flight Research Center, July 2011.

Part 1

Noffsinger: Can you give me your name and your title?

McTigue: My name is John McTigue and I was Program Manager for Dryden Flight Research Center assigned to the Approach and Landing Test [ALT].

Noffsinger: Tell me about what you did?

McTigue: I started on the program just as part time, in 1969, when they were trying to determine what kind of vehicle they were going to have for the shuttle. At that time they were looking at having jet engines on the shuttle for landing and for transporting it across the country. We were doing work with the lifting bodies. We took the X-24B and landed it on the runway to prove that the lifting bodies of those shapes could land on a runway with precision. Based upon these demonstrations the engines were discarded from the shuttle and made it a much more capable airplane.

From that I went to working here, putting in the shuttle related facilities. I had people like [Stanley] "Ski" Markey working, for me who made sure that the facilities were going to be put to the standards that FRC would require for future use after the shuttle program was over.

I had other people – I had Bill Andrews working for me. He was responsible for the Shuttle Carrier Aircraft itself, for getting the aerodynamics and the flying characteristics of the vehicle with the shuttle on top of it to be able to [launch] the shuttle [during the ALT] and also for carrying the shuttle across country. Interesting thing is, you have two different positions on the vehicle, the 747 for carrying the shuttle and for when you are going to have it launch for an Approach and Landing Test. It is raised up at a higher angle for an approach and landing test so that when the airplane, the 747, would push over after it got to altitude that gave a ½ a degree difference in lift between the shuttle and the 747 carrier aircraft so it had a positive separation capabilities and it would separate cleanly and was sufficient distance so the 747 could turn away and allow the shuttle to continue on it's path on down for re-entry, or for entry I should say, and landing at the lakebed.

Once the truck and orbiter reached the main base at Edwards the pair moved to the taxiway where going was easier and unobstructed. The convoy had to stop at the beginning of NASA property so paperwork could be exchanged and the orbiter could be formally handed over. NASA photo

We also got involved here with the landings of the first [orbiter] and for the landing at night coming back from orbit here at the center.

Noffsinger: Tell me about lifting bodies?

McTigue: I was project manager for the lifting body program, the heavy weight lifting bodies. Dale Reed was project manager for the lightweight. The heavyweight lifting bodies were developed by two different groups in NASA; the M2-F2 was designed by Ames, and the HL-10 was designed by Langley. They were built by Northrop Corporation. We actually took the vehicles after they were built and put them up in the wind tunnel to test out their characteristics, at Ames, and they were supposed to be able to tell us about their flying characteristics. They were small enough that we could transport them on the highways.

Noffsinger: It must have caused a lot of attention?

McTigue: Well we covered them. The Air Force had the Martin Corporation back in Baltimore build the X-24A.[32] After the X-24A got delivered here I happened to be on a trip back to Wright Field and was talking with their program managers back there and they had an X-24B shape which had a higher L/D.[33] Now, the basic lifting bodies that we had had a landing profile that was kind of heart shaped, enabling the lifting body to fly to and land within in an area the size of California, not much more than California. But the X-24B had a size that would land in about 2/3 the size of the United States, so it had much

PIGGYBACK SHUTTLE flight testing is nothing new to some retired Center employees. Then US Army Major Jim Love is pictured here in front of an odd combination found after WW 11 in Europe. Results from this type of air launch were questionable then, too.

[32]Eventually Martin Marietta Corp. and then Lockheed Martin Corporation, through a series of acquisitions and consolidations.

[33]Wright Field is now known as Wright Patterson Air Force Base, Dayton, OH.

The orbiter and tow vehicle as it moves down the taxiway at Edwards. NASA photo

more flying capabilities and landing capabilities then what our present vehicles have. I wanted to do the flight test and I convince Paul Bikle we should do that, and we went back to headquarters and Fred Demerrick, who was the manager at headquarters for the lifting bodies, thought it would be a good idea and decided to go in half with what the cost was going to be to modify the X-24A into the X-24B.[34]

Actually, the X-24A sits right within the X-24B. I mean it's still there and it's still within it. It was going to cost a million to modify. I had $400,000 that I had saved aside--I could hide money back in those days, nowadays they make it harder to hide but you can hide it if you are resourceful enough. In any case I had $400,000 and Bikle knew I had it, and so did Fred Demerrick. We decided to transfer that to Wright Field. Well, Air Force headquarters were reluctant to have it done. They didn't necessarily want to have another airplane, but the people at Wright Field wanted to have it done and of course I wanted to have it done. I told Paul it was time to MIPR the money to them.[35] So I MIPR'd the money to them, and I got a call from Air Force headquarters, this guy must have had a few drinks before he called me. He said: "God, what are you trying to do, get me fired?" But that wasn't the case, they decided to go ahead and fund it and that's how we got the X-24B here, just because we funded half of it and the Air Force funded half of it.

Noffsinger: The first time you saw the *Enterprise* what did you think?

McTigue: I saw the *Enterprise* many times. The first time I saw it was in the assembly fixture over at Palmdale, and it was just a tremendous sight to see, a new space ship taking place out there at the Rockwell facility. To me it was a very awe inspiring experience; I was awestruck by the size of it and the capabilities it was going to have. Here you were going to have capabilities to have five or six people on board and the capabilities of working in the cargo bay area. I mean, it was tremendous to see what they were trying to do. Johnson Space Center spent a lot of work and Rockwell has got to be commended for the work they did to build these airplanes that were originally designed to have 100 flights each. There were three of them and there was suppose to be 300 flights they could possibly get out of them, but of course they lost a couple and they built some too. To me it was very inspiring because I learned a lot from Donald "Deke" Slayton. I worked directly under him here. To get the knowledge from the astronauts and all they had gone through to me it was inspiring.

[34]Paul F. Bikle was the center director at the time.

[35]Military Interdepartmental Purchase Request.

Noffsinger: Tell me about the day you saw *Enterprise* roll out, were you there?

McTigue: It was a pretty exciting time period. They had a lot of people there, a lot of press. It was a time to remember, seeing something like that happen. I always felt teamwork or camaraderie because I felt we worked as a team at Dryden. Everybody knew each other's families. You know, you talk about safety here and all the stuff they go through for safety; back when I first came here it was NACA (it wasn't Dryden it was the High Speed Flight Station), we all thought about safety because we knew all the pilots, their wives, you knew their kids. You knew that when they took off and you signed off on that airplane it was your responsibility to make sure they had the best airplane they could possibly have to fly. We took safety very, very seriously. We didn't have check-off lists by the thousands; we didn't go through committee after committee after committee. We just hired people who knew what their job was and let them get their job done. We trusted in people, we didn't trust in procedures.

Frank Brown, Office of Director, and Larry Barnett, Flight Operations, after re-reading the rules about making U-turns on a freeway, bought Public Affairs Officer Ralph Jackson a $30 cigar. . . . .

There was a lot of teamwork on the shuttle program. It was bringing people from all across the country, from Kennedy from Johnson from Marshall together; they had so many people involved that had to work together. Now, in the beginning there was a lot of anxiety that we weren't going to work together, but when people got to know each other, and could trust each other, now that is when the work began. That's when the work got done when people started to trust each other.

Part 2

Peebles: Can you describe your involvement with the ALT flight test?

McTigue: Okay, well my involvement started back in a lot of ways. "Ski" Markey and I came up with the way of getting a shuttle here to begin with. We came up with the cross-country roadway up 10th St. East, which is now called Challenger Way, then through the West Gate, which you couldn't come through now the way they got it. They got it all barricaded, but at that time we came through the West Gate and on in to Edwards and on to the flight line.

Peebles: You were taking the shuttle down a city street. Did you have to move telephone poles, stop lights?

McTigue: That's part of what Ski did. He went through the city government and found out what the restrictions were on the streets. We had to move certain telephone poles and light fixtures and stop signs and stuff like that out of the way. Some of them were removed temporarily and put back up after the shuttle went through but most of them were redesigned to be out of the way for the shuttle for the future. There was quite a bit of extra work that had to be done. Actually, making a new roadway down through a long western end of Edwards Air Force Base wherever Challenger Way stopped and there was dirt road that was essentially hard packed and made capable for the trailers and all

---

[36]All but one of the shuttles, *Endeavour*, were moved overland from Palmdale to EdwardsAFB, requiring extensive modifications be made to the roadways each time. By the time *Endeavour* was built the Orbiter Launch Facility had been relocated from Vandenberg AFB to Plant 42, in Palmdale, which not only facilitated loading the shuttle directly onto the SCA, it enabled other orbiters to be cycled through the Palmdale facility for updates.

Enterprise being trucked from its assembly facility at Rockwell International at Plant 42, Palmdale, CA, to Edwards Air Force Base, where Dryden is located. Every orbiter except *Endeavour* was taken to Dryden overland this way. NASA EC79-10345

the heavy equipment to go over it.[36]

Peebles: Where there any particular unusual things that had to be done such as reinforcing manhole covers and things like that?

McTigue: That was all taken care of by the City of Lancaster. They took on the job of making sure the streets were going to be safe for the kind of load going over it.

Peebles: Once each shuttle arrived at Edwards, at Dryden, it was lifted up by the Mate-Demate-Device and then attached to the back of the 747. What were some of the things that were involved with building it? What was your part in it? What were the structural requirements?

McTigue: All the requirements for the Mate Demate Device were established by KSC and then overseen by the contractors that were assigned by JSC. So those communications that had to be established between JSC and KSC, sometimes they were friendly and sometimes they weren't. There were a lot of communications that occurred. Dryden didn't have as much control over some of this that we would of liked to have had.

Peebles: What are some of the unusual aspects of building it, the tolerances, and things like that? It looks like a skyscraper skeleton.

McTigue: Basically that's what it was. It was a very strong steel structure that was put together to be able to lift the shuttle, which weighed quite a bit heavier than anything we had tried to lift before. Along that lines we had to build other things; we took over a hangar up at North Base, and that hangar had to be completely refurbished inside and out with rooms. We had conference rooms. We had office spaces and so forth

[37]NASA took an existing, large, unused hangar at North Base, not far from Dryden's shuttle area, and built an office complex within it to accommodate shuttle related personnel from Dryden, JSC, and KSC that were working on site. It became a building within a building. Prior to this there had been only a handful of trailers located in what is now Area A, the shuttle area, which was insufficient space to accommodate the massive influx of personnel from the three NASA centers, not to mention the various contractors, as the shuttle program geared up for first launch. John Booksha, Oral History Interview with Christian Gelzer, June 2010, NASA Dryden Flight Research Center.

throughout that whole building.[37] I'd like to get that lumber we put in there. That was all season dried lumber that we used to put in to that structure but when it ended up it turned out to be a nice area to have for offices. JSC had their office out there. I had an office out there and so it was something that had to be done before we even started the Approach and Landing Tests. That's where the people made their plans. That's where all the engineers and technicians that had to support the flights and so forth basically did their planning. We had meetings out there at 7:30 every morning with people from the Air Force, KSC, JSC, and worldwide communications. We actually made tests with all the other centers that had to eventually be involved with the shuttle when it went in to orbit. They tested those all out during the Approach and Landing Tests to see if the communications were working.

The orbiter lifted high in the Mate-Demate-Device (MDD) and waiting to be lowered onto the back of the Shuttle Carrier Aircraft (SCA). Only Dryden and Kennedy had MDDs, and only Dryden's was equipped with elevators and certain other additions. NASA EC81-15404

Peebles: So it wasn't just NASA or Edwards, the Air Force, it was the whole group?

McTigue: It was the whole group and Edwards gave us a lot of support and we used a lot of extra work from Edwards Air Force Base over on the hill. We did all the chemical analysis of the components and stuff that were being used; all the fluids and propellants and stuff that we were going to have, were double-checked by the Air Force over at the rocket base.[38]

Peebles: Lifting the shuttle off of the ground and then bringing the 747 underneath it: were there any incidents or problems that occurred, bugs that had to be checked?

McTigue: Just mainly getting used to doing it the first time. The first time you do anything there is always going to be some difficulties. We didn't make enough clearance maybe for the vertical tail to go up and we had to look at what else had to be modified on the structures so that you, maybe clear some extra beams that were put in that weren't needed. There were some things like that but they were very minor compared to the overall work that was done on the shuttle. Ski Markey also monitored quite a bit of that from just looking at it from a safety stand point for Dryden and he could probably tell you more about the basic specs and so forth of that particular Mate-Demate-

---

[38]"Rocket base," often called Rocket Site, and formally known as Leuhman's Ridge, is part of Edwards Air Force Base and has been the test site for rocket engines since the 1950s. It is home for the Air Force Research Laboratory.

Orbiter To Move Monday; Road Closures Noted

Space Shuttle Orbiter OV-101, the Enterprise, will arrive at the Dryden Flight Research Center next Monday, after an overland trip from Rockwell's Palmdale facility where it was built.

The move is scheduled to begin early in the morning and take most of the day. The 150,000 lb craft will be towed across 10th Street East to Avenue E, where it will travel via a new road across Rosamond Dry Lake to Rosamond Boulevard. It will then be towed to Lancaster Boulevard where it will jog across a new road to the taxiway.

Due to the Shuttle move, the flow of traffic on portions of Rosamond and Lancaster Boulevards between 11:30 a.m. and 4 p.m. on Monday will be occasionally blocked. Traffic on Rosamond Blvd. between Forbes and the town of Rosamond will be blocked in both directions from 11:30 a.m. to 2:30 p.m. Between 2:30 p.m. and 3:15 p.m. traffic on Rosamond Blvd. will be blocked from Forbes Avenue east to Lancaster Blvd. Traffic will be stopped from 2:45 to 4 p.m. on Lancaster Blvd. south of Rosamond Blvd. and north of Old Hospital Road.

During the times these routes are closed, Air Force Security Police will direct traffic to alternate routes. For example, the route northward to Highway 58 from the main base will remain open. The south route on Lancaster Blvd. to 120th Street East will be open except for the period from 2:45 to 4 p.m. Rosamond Blvd. from Forbes Avenue west to the town of Rosamond will open after the Shuttle convoy passes Forbes. Then when the convoy turns south on Lancaster Blvd. all of Rosamond Blvd. will be open to traffic in both directions.

All times are estimates and are subject to change based on actual Shuttle movement.

The Orbiter will be carried on a specially designed strongback which is towed by a commercial carrier capable of traveling at 3-5 mph. The Orbiter will have its tailcone on, and the carrier and strongback will have a total of 90 tires to help spread the load. A caravan of approximately 15 vehicles will provide security, service and other required support.

In addition to Rockwell, USAF and NASA security, the California Highway Patrol and both Kern and Los Angeles County Sheriffs will assist in the move.

The general public is encour-

OVERLAND MOVE. This artist's drawing depicts the Shuttle Orbiter enroute from Palmdale to Dryden Flight Research Center. Moving at speeds from 3-5 miles per hour, the 32-mile trip is expected to take approximately 12 hours.

(ORBITER Continued)

aged to watch the Orbiter move from a location along 10th Street East in Lancaster, rather than on Edwards Air Force Base, due to the closure of several main roads on the base.

The Orbiter will spend its first night here in the old X-15 servicing area, where it will undergo acoustic testing. The test consists of running the engines of two F-104's parked close to the Orbiter. Noise levels will then be measured in the payload bay of the Shuttle.

The following day the Orbiter will be towed to the Mate/Demate Device where it will be lifted off the transporter, and then towed to the Weight and Balance Hangar for measurement.

Device than I can because I was not involved in building it directly.

Peebles: I assume that the Mate Demate Device at Kennedy is the same basic design as the one here?

McTigue: Well not only Kennedy but also they made one for the Air Force out at the north coast at Vandenberg [AFB]. Vandenberg had a Mate-Demate-Device put there that was eventually moved to Palmdale, I believe.

Peebles: Now we get in to the part before the shuttle. There was quite a controversy regarding the shuttle design concerning the pop out jet engines.

McTigue: The shuttle design changed daily from the beginning of 1969 when we had people go spend a couple of months back at JSC. Actually, I think they went back to Langley first because that was where most of the design work was being done to begin with; then it was transferred to JSC. We had people involved back there trying to give them our [advice] on how the vehicle should be built based on our work on the X-15 and other X airplanes.[39]

Peebles: Beyond the X-24B, what was the contribution of the lifting bodies to the shuttle – and I realize of course the shuttle is a winged vehicle as opposed to a lifting body.

McTigue: Actually, on some of the original designs of the shuttle, and as I was saying it was changing daily, there were some lifting body shapes that both Ames and Langley were proposing for the shuttle, but they were put aside because JSC wanted to have a winged vehicle and that's why it turned out to be a winged vehicle. Originally it turned out to be a straight-winged vehicle but that was quickly eliminated due to the problems of trying to keep the heat down on a straight-wing vehicle and coming up with a delta-wing vehicle, which obviously has worked very well.

Peebles: The X-15's significant contributions to the shuttle?

McTigue: The X-15's contributions were basically that we could fly in outer space. We proved that with Walker flying up over 350,000 ft. and we made quite a few astronauts with the X-15, especially the X-15 #3, which was my airplane by the way: I was an OPS engineer on that one [serial number 66672]. Among the things that we learned on the X-15 which we suggested to JSC, was the atmosphere for the pilots during their take off and their landing be a nitrogen atmosphere like we had in the X-15--or at least a normal atmosphere and not a high-level oxygen atmosphere, which resulted in a fire, which was disastrous.[40]

Peebles: The X-15 had a very specific landing profile: the high key energy management and the rest of that. How did that influence the shuttle?

McTigue: Well, that landing profile was very similar to some of stuff we did with the lifting bodies. The lifting bodies also had a very sharp descent and final pull up and landing. We actually had a test airplane that we used for the pilot to become familiar with, and many flights were done prior to flying in the shuttle. It was called a shuttle flight test airplane.

Peebles: [Joe] Engle was one of the X-15 pilots and one of the shuttle astronauts. Did he ever make any comments about the differences or the similarities?

McTigue: I can't remember any of his direct comments but I do know that he appreciated the testing he had done in the X-15 and how it influenced how we flew the shuttle. You would have to ask Joe directly on that to get his comment.

Peebles: Looking back you know the good times, the bad times what stands out?

McTigue: Actually being able to accomplish the complexity of that vehicle and get-

[39]For more information see Peter Merlin, "The Contributions of Lifting Bodies," in *NASA Dryden's Contributions to the Shuttle Program,* edited by Christian Gelzer,

[40]On 27, January 1967, the prime crew of Apollo 1, Lt. Colonel Virgil Ivan "Gus" Grissom (USAF), command pilot; Lt. Colonel Edward Higgins White, II (USAF), senior pilot; and Lt. Commander Roger Bruce Chaffee (USN), pilot, died when an electrical short led to a fire inside the capsule atop the Saturn V rocket. They were training for the flight and the capsule was pressurized with pure oxygen at 15 psi.

ting all the systems and all the people to work together. It was kind of hard to get all the people actually to work together, because each of them had their own way of doing things. JSC and KSC would want to do things certain ways, and FRC wanted to do them other ways, so there were communication problems right from the beginning between the three centers. But they were basically overcome with our Monday morning meeting, or I should say, our every morning meeting at 7:30. The Air Force was there at that and all the centers that were involved in that meeting every day. The schedule was made for the day, was printed and was out to everybody's desk by 8:30 every morning so everybody knew what was going on based upon the communications that were established from that morning meeting.

Peebles: That was an important aspect of getting everybody to play together?

McTigue: It absolutely was. I learned a lot from [Donald] "Deke" Slayton. He was a tremendous guy. He'd have people come to meetings and if they couldn't answer something he'd say: "tomorrow, have the guy that can answer be here." If you were at this meeting you'd better be able to answer the questions that are coming up or have the person you have go to get the answer to be here instead of you tomorrow, and I thought that was a very good philosophy. It keeps communications to a minimum.

Peebles: Looking back, what will you miss about it?

McTigue: Being involved with the people, especially our people. We made a team that talked to each other. We had people who didn't actually go so much by procedures as they did by having people who knew what to do to get the job done. That was a little different then what the shuttle people did. They had procedures for everything; even how to blow up [inflate] the tires on the shuttle. There were a lot of procedures written; thousands and thousands of them for very minor things that a normal mechanic could take care of by himself.

Peebles: Of course, when you're going into space there are no minor things.

McTigue: Well, we were talking about the Approach and Landing Test to begin with, and that was all here on Edwards Air Force Base. Granted, they had to develop procedures that we had to carry over to them when they went in to space, but again, how to blow the tires up is something that is straightforward; there's a normal guy at a tire shop that takes care of them.

Peebles: Looking back, do you think there were things that could have been done differently or better?

McTigue: Well, looking back you can always second-guess people but when you're on the spot trying to make decisions it's a lot different. You made decisions and base them on the people that were there, telling you what was going on and what you basically know. Sometimes what you came up with was the right answer and sometimes it wasn't. That's not the fault of the people; they based it on what they knew and what they had in front of them. It's kind of hard to second-guess people after the fact.

Peebles: There's the joke about when you're up to your rear end in alligators it's a little hard to remember that your original goal was to drain the swamp.

McTigue: That's absolutely correct! That's a saying that has gone on for quite a few years.

Peebles: Anything I've missed?

McTigue: Well, I think the enormity of the crowds that came to see the shuttle land was something we didn't anticipate. We didn't anticipate so many people being involved and interested in it. We had a lot of space set aside but we added more space, as you know, to the west of us, here up on the hill, that wasn't there for the first flight. That was one of the things.

I think inspection of the shuttle in space was something we didn't think about, that came about because of accidents.

Peebles: From a technological point of view could you draw the line between the early X-planes, the X-15, the lifting body, and the shuttle?

McTigue: Well, you have to look at what the early airplanes were designed to do. The first one, the X-1, was designed basically as a 50-calibur bullet that would break the sonic barrier. They knew a 50-calibur bullet would break the sonic barrier because you heard it every time the vets were out here in the range. I heard them in the Marines when they were firing them. So you heard the sonic boom way earlier then we did finally when the airplanes did them. Then the next series of airplanes extended the speed range. We were basically looking at the speed range and the handling qualities. The Phase II's went to Mach 2.

The X-15s obviously went on up to Mach 6+ and altitudes above 350,000 ft. The first X-planes didn't go over 80,000 ft. or so, so there was a big difference technology wise. The X-15 had to have a pressurized cockpit for the pilot. The other airplanes didn't because they didn't go that high.[41] The shuttle had to go out in to space, which was basically a vacuum, and so the structure and the building of airplane had to be very precise so we didn't have leaks. We had all kinds of leaks in the X-15. I was constantly repairing leaks on the X-15's cockpit to keep the pressurization right. They did that all in testing at different stages of building the shuttle so that when they finally went in to space they had a very tight airplane that wasn't leaking.

The development of the engines was an entirely different thing. The X-15 engines were developed over hard lessons. We had lots of failures with the X-15 engines. The shuttle engines, when they were finally developed, turned out to be excellent. They haven't had any problems with the shuttle engines through all their flights. Rockwell did a tremendous job in developing the rocket engines for the shuttle.

It's an entirely different world now. Everything now is digital; back then, when we first started with the shuttle it wasn't all digital-it had a lot of analog work. But it's progressed and they've got the glass cockpit now that they didn't have at the beginning.

Peebles: When I do the tours I tell people that when the X-15 project started there were biomedical experts who weren't sure a person could function in weightlessness for two whole minutes.

McTigue: That's correct. There were a lot of medical people involved right from the beginning on the X-15 and their thoughts were that the pilots weren't going to be able to take the stress of the accelerations, the pilots were not going to be able to take sitting in that environment for any length of time. They were proved wrong but they were proved wrong by a lot of discussions and unfortunately people were coming up with

[41]In the 1950s pilots who flew above 50,000ft invariably wore partial-pressure suits; by the time of the X-15 they were wearing full-pressure suits, virtually the same suits as the Mercury astronauts.

ideas that had been proven wrong by other people who had been to altitude by balloons and so forth. It was hard to convince them and they had voices in Congress that made us do extra things that we didn't want to do but we had to in the long run. We got the job done.

Peebles: I want to talk a little bit more about the uniqueness of the MDD. I've heard that it was designed to be portable . . .

McTigue: Well, the first design they talked about was only making one of them and moving it from site to site, which was a [stupid] decision. The one that we have is definitely not a portable one. People have ideas and portable was one of them. I think they have a portable one, at Palmdale don't they?[42]

Peebles: That has been removed.

McTigue: Oh, it has. Okay.

Peebles: As the shuttle is in the MDD and the SCA gets involved, can you walk me through maybe a day in the life of working that scene?

McTigue: Well, you go through a lot of time in the Mate-Demate-Device getting it ready, testing it, making sure it is structurally tested before you do the lift. They had heavy weights that they lifted and tested everything out with. They tested all the instrumentation on it to be sure that they were recording all the weights and they didn't have anything off balance some place, and then on the day of the lifting they had inspectors at all levels looking at the mechanisms as [the orbiter] was being raised, making sure nothing was going wrong. That was all prearranged and people knew what their job

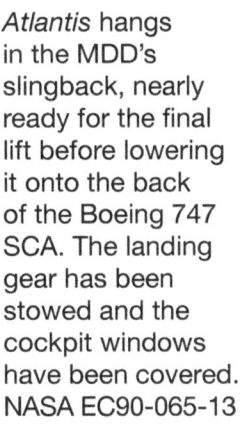

Atlantis hangs in the MDD's slingback, nearly ready for the final lift before lowering it onto the back of the Boeing 747 SCA. The landing gear has been stowed and the cockpit windows have been covered. NASA EC90-065-13

[42]The Orbiter Launch Facility, or OLF.

was.

They took a lot of time lifting that vehicle safely in a very precise manner. Then they [pulled] the 747 in, and set the shuttle back on top of the 747. After they checked it out and it was mated they were able to remove the 747 and the shuttle from underneath the Mate-Demate-Device and do the final check out of the shuttle and preparation for its flight on top of the 747. That usually took about a good day or day and a half just to do the final checkouts.

Peebles: We live in the Mojave Desert and it has two weather settings: too hot and too cold. What are the maintenance requirements of the MDD? In other words it could be a 100 degrees or snowing.

McTigue: It was basically made for all the environments it was going to see.

Peebles: Let's go back in time. You're coming in to the first lift. Give me the details of what you're experiencing?

McTigue: Anxiety! First, getting all the systems ready which were not ready when the vehicle [Columbia] first arrived here. A lot was done on the basic control systems for the shuttle. It was done here before it was actually raised up in to the Mate-Demate-Device. Going into that first lift I was very anxious. Is it all going to work? Had we done everything right? Are we going to have a cable fail or is something going to slip? It was all those things going through your mind even though you know you've checked them 50 times, in the back of your mind something is going to happen. Fortunately nothing happened. We went through and I left smiling because it really turned out to be a very good operation and all the testing we did proved that we could make it happen.

Robert R. Meyer, Jr.

Robert R. Meyer, Jr., spent most of his aeronautical engineering career at NASA Dryden, although he participated in a personnel exchange program at Langley where he worked with Richard T. Whictcomb on transonic wind tunnel research. His first experience at Dryden was as a co-op student while still enrolled at Purdue University, and his first project was the Ground Research Vehicle Drag Reduction Program dubbed the Shoebox.[43] During his career he worked on aerodynamic loads testing of the shuttle's thermal protection system, the F-18 High Angle of Attack Research Vehicle (HARV), and aerodynamic flow on an F-111, and F-14, and an F-15, among other things. He flew as flight engineer on the center's SR-71s in the 1990s, and has held various positions in management, including serving as deputy director of the center. At the time of the interview he served as the program manager at Dryden for the Stratospheric Observatory for Infrared Astronomy (SOFIA).

Bob Meyer, interviewed by Guy T. Noffsinger, Shuttle Documentary Interviews, October 2010.

Noffsinger: If you could give us your name and your title then go into the first story?

Meyer: Bob Meyer and I am the program manager for the SOFIA Program here at Dryden.

Noffsinger: Tell us about the first story; I think it was about the crowds.

Meyer: For the first shuttle landing there were a lot of people that wanted to come out to Dryden and to Edwards to see the landing. So what the center decided to do was have the employees here serve as tour guides for that event. We got to see a lot of celebrities that came out. As I recall, Clint Eastwood came out and then Jerry Brown, the governor, came out. A lot of Star Trek actors came out, John Denver the singer came out. So we were all pressed into service to ride in the bus. There were usually two of us in the bus that rode out from Lancaster as they were brought out to come see the landing. Basically, they didn't bring in professional tour guides or anything we basically were tour guides for all of that.

Noffsinger: Tell us about the load test?

Meyer: As it turned out, the first time the orbiters actually flew was on top of the 747 Shuttle Carrier Aircraft. We learned a lot was from those flights.

Columbia, when it was first taken down from Palmdale or Southern California to the Cape Canaveral, experienced some tiles that came off the orbiter itself, and some came loose, and so there got be a real concern with the program--if they were coming loose during this ferry flight down to Florida, what was going to happen during an actual launch. So a group of engineers from Johnson [Space Center] came out and said: "We want to do some more work with the shuttle tiles and the air-loads and we would like to know if we could fly some of these on the Dryden aircraft because these are going to

---

[43]See Christian Gelzer, *Fairing Well, From Shoebox to Bat Truck and Beyond: Aerodynamic Truck Research at NASA's Dryden Flight Research Center*, (Washington, D.C.: NASA Monographs in Aerospace History #46, 2011).

be really big test articles. They won't necessarily fit in a wind tunnel very well and could you guys help us?"

Not too long after that we did a rain erosion program, and again, that was because they got a little too close to some rain [on a ferry flight]. This was pretty early in my career. I served as the principal investigator and the chief engineer for that project. We ended up flying tiles on our F-15 and an F-104, from six locations on the shuttle that were deemed to be subject to fairly high loads or were critical. We flew the thermal protection system tiles that represented areas like the window post around the windows for the crew, the leading edge (they were called J-tiles), we flew them, we flew the leading edge of the vertical tail, and some trailing edge tiles. There was a tile called the "close out" that closed the area out from behind the reinforced carbon-carbon and transitioned into acreage tiles. We flew all these tiles and actually simulated, not in time but in air-loads, the loads they would see during a launch on the shuttle. We actually flew them to 1.4 times that of the load they would see during a shuttle launch. We actually over tested them. And it turns out that some of the tiles that we flew had very low margins and had to be beefed up or strengthened before the actual shuttle flight STS-1.

We finished the air-loads test and then later, on another flight down to Kennedy with the orbiter itself on top of the shuttle carrier aircraft, they got a little bit too close to some thunderstorms and flew through the edge of some rain with the orbiter. When they landed at Kennedy they noticed that some of the tiles had been damaged by the rain itself. That prompted yet another test project to actually try to quantify what damage flying through rain with the vehicle would cause. So we did some testing out here, and part of the reason the test came out here was that we had the relationships with the engineers at Johnson from the earlier air-loads tests. But there was also a tanker that was set up actually for icing tests here at Edwards. It was essentially a refueling tanker that had this big ring on the back of it that would spray water out and for normal icing tests it would turn into ice, but it could also be used for rain. It could create rain out of this ring. So the idea was that we were going to fly behind this ring and allow the water that came out of it to impact the tiles and see what damage was caused. We started the tests and we had a fixture on the bottom of an F-104 that had the thermal protection system tiles mounted at different angles so that when the rain hit it we could simulate the tiles that were

SCW Inventor Receives Award

Dr. Richard T. Whitcomb, inventor of the NASA Supercritical Wing that was first flown by the Flight Research Center, received a $25,000 cash award from NASA for his invention.

This is the largest cash award given by NASA to an individual and was presented at the recommendation of the NASA Inventions and Contributions Board. Several Center personnel participated in the evaluation of the concept.

Flight research conducted here demonstrated that the new wing shape allowed the aircraft to operate more efficiently.

NASA's NB-52B sits on the taxiway following a minor mishap: the rear hook holding the "blivit" in its pylon gave way followed by a thud. The scene was visible from the Dryden pilots' office where Bob Meyer was at the time. A flatbed truck was pulled up with a cradle to lift the shuttle parachute test device off the taxiway. NASA EC83-22789

parallel to the rain or that were perpendicular to it. The tanker put out as much water as it could and went as fast as it could and there was no damage to any of the tiles. We were scratching our heads trying to figure out what happened, and I was kind of seeing my career flash before me because, here we designed this real expensive experiment and it wasn't working. It turns out that, not unlike this week here in the desert, we had a rainstorm come in and we actually had the tanker schedule but we cancelled it. We said: "well, we are going to go out and fly in some real rain because we've got it." The instant we hit rain the tiles almost exploded. So of course I was very relieved because the experiment, the text fixture project we put together, worked. But we were curious why the tanker here at Edwards hadn't worked, and actually kind of a spin off, kind of a side thing from the project, we found that the tanker didn't work correctly. What was happening when the water was coming out of the nozzle on this ring was that the air-flow was basically sheering the water, and there were no droplets coming out, just mist so it wasn't actually representing rain correctly. After that they shut the use of the ring, or that water spray, down.

Those tests helped the operational team establish some of the launch and landing limits for when the shuttle took off and when the orbiter came back in. We did some work on the Return To Flight and we did drag chute qualifications here with the B-52, so before the drag chute flew on the shuttle it was qualified on the B-52.[44]

One of the other projects that Dryden supported was the Solid Rocket Booster recovery parachutes, and we had what we all lovingly referred to as a blivit that represented

[44]The drag parachute was initially part of the orbiter but was dropped from the list of flight components before the launch of *Columbia*. Following the loss of *Challenger* and the down time the drag chute was revived and became a permanent feature of the orbiter, used to help slow the shuttle after touchdown.

the mass of the solid rocket booster after it had burned out.[45] It would be carried up under the wing of the B-52 and dropped and then we would test the recovery system. One day I remember being in the office and looking out the window and the B-52 was about half way down the taxi way toward the main runway at Edwards; the blivit was attached to the front of the wing but the back of it was laying on the taxi way; the hooks for it had broken. There had been some fatigue on the hooks that held the back of the blivit and it had broken the hook and had actually fallen down and was lying on the runway. Fortunately that didn't happen in flight while we were trying to do a flight test with the system, it happened while we were taxing out, but it was pretty scary.

Noffsinger: Tell us anything about the first landing, about what the crowds were like, the excitement, all the commotion it caused. I heard a story that people wanted to go out to the orbiter to greet it. The National Guard had to scramble their helicopter in circles.

Meyer: The first landing was pretty exciting, and for me personally it was real exciting because of course we had done these air-loads tests on the tiles before it entered, and your're wondering: "Okay, did my work really work correctly? Did I miss anything?" Of course, the orbiter came in and landed just fine. There were hundreds of thousands of people here. The crowd has been estimated at 250,000-300,000.[46] They had people on the far side of the lakebed that could drive their cars in and could park out there, and then they had the VIPs that were on the west side of the lakebed, and we had bleachers and so forth and, we had to bring in the National Guard with helicopters and they actually had guns on board the helicopters and they would fly back and forth in front of the crowd over on the east side of the lakebed for crowd control because people wanted to run out and go greet the orbiter and of course they couldn't do that.

[45]One of the definitions for a blivit is 'something that is hard to describe.'

[46]Later estimates put the crowd at between 400,000 and 500,000; people had gathered on the eastern shore of the Rogers Dry Lake for several days in anticipation of the shuttle's landing.

Pete Seidl

Pete Seidl began working with aircraft in 1974, at Vance Air Force Base, Enid, OK. From there he moved Sheppard Air Force Base, Wichita Falls, TX, working on T-37s and T-38s. With that experience he took work with NASA at the Johnson Space Center where he worked on T-38s. He left NASA to work on E-4Bs; in 1979 he was offered the position as site supervisor for the 747 Shuttle Carrier Aircraft (SCA). He spent the next 29 years in that position and retired January 28, 2008.

Pete Seidl, interviewed by Christian Gelzer, June 13, 2011, NASA Dryden Flight Research Center, Shuttle History Documentary Interviews, Edwards, CA.

Gelzer: It's June 13 in the afternoon and we're talking with . . .

Seidl: I'm Pete Seidl. I started my aircraft career back in '64. I was on the wash rack at Vance Air Force Base, washing airplanes. I was there for about five years and moved on to Sheppard taking care of G-2s and T-38s at Sheppard for about five years. The reason it's every five years is [that] the contracts usually run about five years and as soon as they're up you usually get a new contractor, so I moved a lot. I went to Houston and worked with NASA on T-37s and T-38s. I had a great time. All the Apollo stuff was going on and all that stuff was still in action. It was a fun time. After that I moved and spent two years at Greenville, Texas, working on E-4Bs, modifying them and updating new ones.[47]

Gelzer: Do you remember meeting Gordon Fullerton when you were at JSC?

Seidl: Oh yes! Gordon and me knew each other. As a matter of fact, when he got to town [Lancaster] he asked me and I got him a house right next door to me so he lived right next door to me for several months. What a good guy; what a nice neighbor.

Gelzer: When they brought you on here at Dryden for the Boeing 747 what position did you come to fill?

Seidl: Actually I filled a mechanics position for about two years and Jerry Eddy was the manager or site supervisor and then he was kind of… he had a wild life. He liked to enjoy things. He was just a character. Everybody loved him. So anyway, no fault of his, but he had to leave the state. (Laughing) After that I took over, and that was '81.

Gelzer: Tell me about your job?

Seidl: My job; I was a site manager or site supervisor for the 747 Shuttle Carrier Aircraft and what that pertained to. We did all the aircraft maintenance for on the shuttle carrier. We did that for a long time without computers, but it got kind of hard after the computers came in. It used to be a lot more fun then it ended up being. We had good times back then.

---

[47]The E-4B is a Boeing 747-200 modified to serve for the Air Force as a "National Airborne Operations Center for the president, secretary of defense and Joint Chiefs of Staff." http://www.af.mil/information/factsheets/factsheet.asp?id=99, accessed December 16, 2011.

Gelzer: When you came on in '79 did they already have two airplanes or was it still one?

Seidl: No, just one airplane. The second airplane didn't come until after the *Challenger* accident. The Rogers Commission decided they were going to do away with all the single point failures and the SCA was one of those single point failure potentials and that's how we got that, years later.[48] Let me see if I'm right here. I think the *Challenger* was '86 and then we got the airplane in '88. That's when we got 911.[49]

Gelzer: When you came, the Approach and Landing Test program was already over with but *Enterprise* was still here physically, or had *Enterprise* been moved some place else?

Seidl: *Enterprise* went around. We would actually take it different places. It wasn't out here. I think it was at Palmdale but they actually moved it around, like I say, if somebody wanted it--if KSC (Kennedy Space Center)--wanted it, they would take it to KSC. As a matter of fact, on my trips several times we went over to Vandenberg AFB.

Vandenberg was supposed to come up with a shuttle launch facility. Well, that ended after the *Challenger*. They had filament-wound boosters that were lighter and supposedly they could blast off and do their polar orbits. After *Challenger* they decided it was too risky to use these filament-wound boosters because they might blow up on them or something. So they scrubbed that whole thing. They spent billions of dollars on that over there.

We went over there with the *Enterprise* several times and did fit checks with them and where they were going to do their Mate-Demate-Device and we did an awful lot of work over there, and then they scrubbed it. That was really too bad.

Gelzer: You were here for the very first STS-1 landing and all the ones subsequent to that until you retired so?

Seidl: Right. You know, I have a lot of good memories. I guess the early times were the best. It used to be a lot of fun. There used to be no pressure. They didn't tell you there was a million dollar's a day to move the orbiter or any of this so it used to be fun; then they started telling you this, it became a job. It was a lot of fun. Of course we were almost living in Florida when we first started the launches and I spent 50% of my time in Florida, just about.

Gelzer: Doing what?

Seidl: Taking the orbiters back after landings because everyone landed out here then. So we were taking orbiters back. As soon as one would land we would take it back. Of course back then the turn-a-rounds were like three days or so.

Gelzer: Once it was here it would take three days to turn the orbiter and send it home?

Seidl: Yeah send it home, three or four. At the end it ended up being like seven or eight

[48]Report of the Presidential Commission on the Space Shuttle *Challenger* Accident, In compliance with Executive Order 12546 of February 3, 1986, chaired by former Secretary of State William P. Rogers.

[49]"911" is a contraction of the airplane's full registration: N911NA. The first Boeing 747 SCA had tail number N905NA. The first digit in the "N" number of NASA-registered aircraft indicates which center is responsible for the aircraft, in this case, the Johnson Space Center. Dryden's identifying number is 8.

days and that's all due to safety concerns and all this stuff. At first it was a lot looser. I don't think they endangered anybody but they just did their jobs. I don't think they had the oversight like they had at the end.

Gelzer: Once the orbiter is down now there are quality assurance people that go around and monitor the tasks that the people are working after the orbiter has landed.

Seidl: Everything! It really changed. Yeah it really did. Of course they had a guy by the name of Bobby Horn from Florida there that would run the operations here. He was a really, really sharp guy. He has passed away but he used to really work those guys and they'd turn those orbiters around, but all of that changed after *Challenger*: it seemed like it brought a whole new regime down. It got a lot harder.

Gelzer: In the early program, when the orbiters were still landing on the lakebed, did you physically go out with some of the other Dryden people to work with the orbiter or did you stay back at the shuttle Area A?

Seidl: We usually stayed back. We'd usually go out and watch the landing then come back. We didn't have too much to do with the orbiter. We just waited until they got it ready for us and then we'd mate it [to the SCA].

Gelzer: You became a player once the orbiter was ready to be hoisted.

Seidl: Yeah! Once it got into Area A, we were players.[50] Of course, every time they came in we'd have to move our airplane to make room for it to go by the back end. We got pretty busy. What was really hard on turn-a-rounds and stuff were the late nights. I had five people most of the time. We got up to eight people when it got really, really busy, but a lot of the times we had five people and of course they run three shifts and we have to support them so they would be out there at midnight wanting power on the airplane so they could check their TR units and make sure the orbiter was running and had all the heaters that worked and all this stuff. Well, we were dog-tired. We had people spend the night in the cockpit usually and if it was cold, even in the orbiter when the orbiter was in the MDD we had to keep the power on. It was a long drawn out day – nights I should say – days and nights.

Gelzer: You were one of the people that had to hustle off to White Sands Missile Range, New Mexico, for STS-3?

Seidl: Actually, I kind of lucked out on that one. I went to Florida, enjoyed the time off, me and John Goleno went on to Florida and the rest of the guys had to do the work on the airplane there.

Gelzer: Did you go back to White Sands for the trip home from Florida?

Seidl: No, I didn't even go to White Sands. There were four other guys there and they covered it. We waited for them to come into X-68, which was in Florida.[51] We were

[50]The most northern area of the Dryden facility, Area A, was reserved for shuttle operations. This was where the Mate-Demate-Device stood and where all the ground support equipment was stored. When a shuttle landed a gate with a guard effectively closed off Area A to only those working on the orbiter and those with necessary access to the area.

[51]KSC's Shuttle Landing Facility (SLF), a single runway, carries the FAA identification code X68.

waiting there. We would get all the logistics set up, the tug and the stair trucks and all of that stuff ready for them. That was our job that time, so I really lucked out because they said it was really miserable. The wind was blowing, the gypsum was all over--it was really a mess. They were finding stuff in this equipment for years after that. They would take something apart and the gypsum would be inside of it.

Gelzer: The year you came, or was it the next year that NASA sent *Enterprise* to the Paris Air Show?

Seidl: That was a real fun trip. Before we even went they decided that they needed a missile defense put on our aircraft, so they had a team from Boeing come out; it took them about a week and they installed an infrared missile defense. It was quite a job. Back then it was supposedly a secret. They told us: "don't tell anybody, don't tell anybody what they are." So we kept it a secret; we were trying to. After we got to Paris Air Show all of the refuelers were asking us what those were. We had to tell them they were directional lights for in-flight refueling. We didn't even have in-flight refueling on the airplane so that was a farce, but we went along with it. We left here and had to go way up north and across to get over there because we didn't have very long range with the orbiter on the back, 1,500 miles I think was probably our max – pretty short legs.

Gelzer: So you went Newfoundland, Canada, Iceland--

Seidl: First we went to Colorado Springs, Colorado– I guess this was a show-and-tell before we started our trip, and then to Wichita, Kansas, Wright Patterson Air Force

The SCA with *Enterprise* on its back over Edwards, CA, as it begins its long trip to the Paris Air Show. In the background is Rogers Dry Lake.

In preparation for taking the shuttle *Enterprise* to the Paris Air Show in 1983, NASA put a type of anti-missile device on each engine pylon of the Boeing 747 SCA. When asked about the devices at the show, aircraft personnel offered different answers, including that they were used to aid in-flight refueling. NASA EC83-24285

Base, Ohio, Goose Bay, Labrador, then to Keflavik, Iceland, which is really a fun place: they said there was a woman for every tree--of course they didn't have any trees.

Gelzer: How long did you stay at these spots?

Seidl: We didn't stay very long. We overnighted at a lot of them but some we didn't. Some of them were just a gas-and-go. Keflavik, Iceland, was a gas-and-go, but Fairford, England, was the next stop. These people were crazy. NASA sent a whole bunch of PR stuff with us and they gave us these little pins, shuttle pins. As a matter of fact I got one here. Anyway these pins were like gold. The English thought they were worth a Bobby's uniform.[52] When we went out of there they gave us five or six full uniforms, the hats, the pants, the shirts. Most of them had to go home and said: "I'll come back; I'll come back and give you those." It was quite a show. They were great people.

England was really a nice place. After that we went to Bonn, Germany. We had a good time there. They really treated us nice in Bonn. They went overboard. Gosh, we went bar hopping that night and couldn't buy a drink, those people were so friendly. Shots for everybody; it was a fun time but we all felt like hell the next day.

Then after that we made it in to Paris. In Paris we were just supposed to display the airplane. There weren't supposed to be any flights at all in the show. Well, Joe Algranti was in charge. He was the director of operations at Houston [JSC], flight operations. Anyway, he decided he wanted to fly it, so we started flying. We flew three flights on the 27th, 28th, 29th of May. The people just loved the flights but they hated us because

[52]"Bobby" is the vernacular for British policeman.

it was such a cramped area in the parking space where we were. These vendors would have to move all their little tents and everything when we went by, so we were very unpopular. They got over it finally but it was pretty trying.

Gelzer: What was the perception in terms of the attendees at the Paris Air Show?

Seidl: It was outstanding. Whenever we flew there were people who would follow the airplane. When we were taxiing they would just be walking beside it. It was just unreal. I don't know how they got all those people in there but they did.

Gelzer: Did you have to work the airplane? That is, were people coming in to the airplane to look at it or just staying on the ground?

New hours for tours of the Dryden Flight Research Center have been established. They are weekdays, 10 a.m. to noon, and 1 p.m. to 3 p.m. Tours begin at the Visitor's Center in the Integrated Support Facility and feature a film, briefing, walk through the Center hangars and a tour of the Space Shuttle facilities. For large groups (10 or more) reservations should be made by calling Glenn Briggs, ext. 221.

Seidl: Actually, we were over on the airfield part. We didn't have too many tours there. We did have some, the VIPs, but the public couldn't get in it. I think that was a security concern. We had a lot of VIPs. What was funny in Paris were the maintenance people: they wouldn't let us drive their tug--because of insurance reasons they told us. We said: "Oh, okay, but we would like to tow it ourselves." They said: "no, you can't do it - insurance." So we let them tow it. The first time they towed it they towed it into the mud. I don't think any of them had towed a big airplane before.

Gelzer: I was going to say I would assume they would need 747 experience to tow that.

Seidl: We thought they had big airplane experience but they turned too sharp and got it in the mud. First time was enough: we said, no, we'd towed it from then on, but they

The crush of people at the Paris Air Show. Here French motorcycle police are on hand to help keep order. NASA S83-36061

really raised heck.

While we were over there, Joe Algranti decided we'd stay downtown. Le Bourget Airport is way over on the northeast side of the town and we were miles and miles away from there. So every day we had to drive from our hotel to the airport. The embassy there gave us some real nice cars. We had black Mercedes Benzes and I thought: "boy, this is going to be nice." But when you start driving in Paris all these little bitty cars are honking and honking. Every time they want something they honk. Anyway, we're driving these black Mercedes trying to get over there and oh, my gosh, it was like a swarm of bees; you couldn't hardly move – the car was big compared to most of the cars over there. It was a nightmare. Of course, I was the designated driver.

They were having student riots while we were over there. Our hotel was Hotel Luxembourg and the Luxembourg Gardens was right by there and it was a big hangout for students. Anyway, they started charging them more – their student fees went up so they were rioting. It was just unreal. These cops would come on their little motorcycles and they had two people on a motorcycle, one of them had a big old whip kind of thing and they would go around after these students. We were going to a bar down the block from there and Tom McMurtry, one of the 747 pilots, got with these students. It was funny--he just disappeared. Pretty soon we were all in the bar there and Tom came in and he stunk to high heaven and I said: "Tom, what do you got on you?" They tear gassed him along with the students. They were after these students and they tear gassed Tom. He came in and then the bar guy got kind of uptight. He was worried about them knowing Americans were in there. He thought maybe some of those students might bother us so he asked us to kindly leave. We went back to our hotel down the block.

Gelzer: How long were you in Paris--for the entire air show?

Seidl: Yes, throughout the entire air show. We left a little early, actually the day before the air show was over, and went to Rome. Rome was just a one-nighter thing. We had to fuel the airplane and get it ready so we were pretty late, so we had this taxi driver that was taking us to our hotel. This taxi driver was a great guy. He was really nice. He decided, seeing as we were just going to be there one night, he would show us around. He took us to the fountain of Trevi I believe it is, and then by the Coliseum. We saw all

The SCA and *Enterprise* while taxiing at the Paris Air Show. One of the crew has popped out of the emergency exit with U.S. and French flags. NASA S83-36062

that. We enjoyed it, and then he showed us a good place to eat closer to the motel. We ate there then walked home--had a great time in Rome.

Gelzer: Was the airplane on display the next day?

Seidl: Not at Rome.

Gelzer: What was the reason for going to Rome?

Seidl: I guess they just requested us to come.

Gelzer: Did you get there during daylight so people could see it?

Seidl: Yes we did. As a matter of fact, you can't fly with the shuttle at night. No night landings with the orbiter.

Gelzer: Who would look after the airplane when it was on the ground at some of these airports like in Rome or in Bonn?

Seidl: Actually they had NASA security. They usually have a NASA contingent, it would be like two or three people, but then the security for the bases we landed at usually guarded the airplane.

Gelzer: Did you always go to a military base?

Seidl: Yes, we always went to a military base.

Gelzer: Except for Paris, you were always at a military base?

Seidl: Actually Le Bourget (Paris) is military too. The military base is where they have the air show: de Gaulle Airport is several miles away. That caused us trouble, too. For the first flight at the Paris Air Show we got a French colonel to fly with us so he could show us how to go and how to stay out of the traffic zones around de Gaulle. This colonel got in but evidently he had a mind of his own and he went too far out. In the paper the next day it said 'NASA breaks periphery of air show'; we got out of their little box, so that was a big stink. It was like we really messed up.

Gelzer: What altitude were you flying at?

Seidl: Well, it was pretty low. They usually fly a couple thousand feet.

Gelzer: Just a couple thousand feet so people can see things pretty clearly . . .

Seidl: Yeah, I think that colonel . . . I don't know if it was in the translation or what, but we got outside of this area. When we got to Italy they said we could fly anywhere we wanted to.

Gelzer: Did you tour the city before you landed?

Seidl: Yeah, fly over? Yeah we sure did.

Gelzer: And I guess the same thing on departure?

The orbiter and SCA flying over Roman ruins on departing Rome. This came after leaving the Paris Air Show for a day. NASA S83-40796

Seidl: Yeah pretty much. We'd fly around and let everybody look. It was a lot of fun, a lot of people. It was just amazing.

Gelzer: Did you run out of pins to give away?

Seidl: No, they gave us thousands of those pins. My gosh, we had so many pins I think we even came home with some. I was really amazed at all the stuff they give away but they had a lot.

Gelzer: And you flew--unlike taking an actual orbiter back to Florida, which required everybody to fly on a Pathfinder--you flew on the 747 with everybody.[53]

[53]With active shuttles returning to the Cape, NASA arranged for a Pathfinder to fly ahead of the SCA by about 100 miles, to look for any kind of weather for the SCA and orbiter to avoid. This was because on the first trip to the Cape with *Columbia* the orbiter not only lost a number of tiles but encountered rain, which caused a shocking amount of damage to the tiles. The Pathfinder sought clear, cloudless skies to fly through and deviated around all other conditions; the SCA followed suit.

Seidl: Yeah, right. As long as the *Enterprise* or one that doesn't have the hypergolic fuels in it you can fly on them, but once they get the hypergolics it becomes hazardous.[54]

Gelzer: So, for the first trip from Palmdale to Florida you could actually fly on the 747?

Seidl: Yes.

Gelzer: From Rome where next?

Seidl: Back to Paris. We went back to Paris and did a fly over for the last day of the air show and then we went to Stansted, England, and then it was back to the pins for uniforms.

Gelzer: How many uniforms did you take home?

Seidl: We had five or six complete.

Gelzer: Each?

Seidl: Each of us had one. Each of us had a set and I wore mine a long time for Halloween, my Bobby hat and my Bobby suit. It was really nice.
   Taking off from Stansted was kind of interesting. We took off, actually we weren't even off, we were taxiing out and we had a hydraulic leak and we said to Joe Algranti: "why don't we just stop and change it before we get airborne?" He said: "no, nope." Joe was always a stickler for schedules. He wanted to be on time and be there, so we took off, could barely get the gear up. We finally got the gear up and headed for Keflavik and we radioed Keflavik that we needed a light-all to change the pump.

Gelzer: What is a light-all?

Seidl: Just a bank of lights that light up the area. That was kind of funny in itself. They probably laughed at us when we ordered them but they had them when we got there. When we got there it was pretty late but we were working changing that pump out at 2 a.m. and it was still daylight, so we never used the light-alls. That was quite a learning curve there. After that we went to Goose Bay, Labrador, that was just a fueling stop, and then on to Ottawa, Canada. Ottawa was a lot of fun. There were a lot of people out--they evidently had the word we were coming in because that whole place was just surrounded with people, it was so busy.

Gelzer: You were the only thing they were coming out to see?

Seidl: Yeah, and we could hardly get out of the base because there were so many; the traffic and stuff. That was kind of fun too. We got there fairly late and the pilots went in to do an interview. We fueled the airplane and we were ready go. We came out and here's this big limousine and a van. The limousine was for the pilots and we knew it. All the maintenance crew decided we would take the limo, so we took it and left them the van.

Gelzer: Did you hear about it later?

[54]Rocket propellants that combust spontaneously when they come into contact with each other, such as nitric acid and hydrazine, are referred to as hypergolic. Hypergolic mixtures are invariably toxic.

Seidl: No, no they didn't even know there was a limo there. I figured I'd hear about it but I never did. Take off from Ottawa was fun. They had so many people there all down both sides of the runway, and at the end of it they even had people across it, in front of us--if we'd of run over them we'd of killed hundreds of people. Ottawa has a runway that dips down and goes up. We took off, put the power to it and away we go. I was up in the cockpit looking out of the window and you see all these people along the sides, and at the end I could see these people and there was terror in their eyes. Their eyes were like silver dollars because we were probably 50 to 100 feet above them at the end of the runway and here all these people were under the airplane. I'm surprised we just didn't blow them away with our wash.

Gelzer: Well, you don't know if you did or not, actually.

Seidl: They never said anything so I guess it was okay. I was looking out of the window and you could see their eyes. After Ottawa we went to St. Louis, Missouri, just for a gas stop, then went on to Dulles, Virginia. That's where they gave us a public service medal.

Gelzer: How long were you in Dulles, because I'm sure that's where you showed the airplane again.

Seidl: Yeah, we did, that was overnight. We stayed overnight there and then we went to Sheppard AFB after that, on the way home. That was our next stop. We stopped at Sheppard, where we got these contract fuel trucks and they had a hose, probably 2" hose; they contracted to get the fuel and I have no idea why we didn't use the base fuel, but we had this contract fuel and it took us all night to refuel just about because it was slow. We had two trucks and they were switching back and forth running back getting it, running back and getting it.

Gelzer: You guys never left the airplane basically.

Seidl: No, there were a couple guys I think that had to stay. They stayed until pretty late. Then, 13th of June, we finally made it home; that was one of the best trips we ever took, I think. What really makes it nice is that we had the *Enterprise* so you didn't have all the purge requirements and all this stuff. It was really nice.

Gelzer: Once you came back did they take off the missile defense?

Seidl: Yes, as soon as we got back they took all that off. It wasn't on there very long. It went by the way side. Never put it on again. Still have the hookups and everything but never had any use for it; I guess that was the last big hoorah. Too bad they don't do something like at the end of the program here.

Gelzer: Flying it around?

Seidl: Yes, actually they could pick up *Enterprise* and drop off another orbiter there at Dulles.

Gelzer: Tell me about your trip to Mobile, Alabama, with *Enterprise*.

Seidl: That was for The World's Fair, yep, in St. Louis. We off loaded *Enterprise* in Mobile and they barged it to The World's Fair. After that, after the show was done, they put it back on the barge and then went around to X-68 and dropped it off over there on the barge.

Gelzer: How did they take the *Enterprise* off the 747 in Mobile?

Seidl: Well, you know they have the recovery ships and everything for the boosters so they have a great port there so they just used the same place.

Gelzer: Yeah, but how did they take it off the airplane?

Seidl: With cranes, portable cranes.

Gelzer: They brought in some cranes and the actually did like they would have had like they did at STS-3.

Seidl: Like they did at White Sands.

Gelzer: Then how did they get *Enterprise* to Dulles?

Seidl: We picked it up at Florida again, X-68, and took it over to Dulles and that was just a one-hop thing. We left it at the Smithsonian there and there they had the portable cranes.[55] I think Nicholas Cranes, they do all the lifting, they were there. It was kind of a fun trip. I had a lot of relatives there and they came out and looked at it. They were really amazed at all the big fittings and all the stuff they use to pick up that orbiter. They were just amazed. I was glad they got to see because my uncle passed away not

In a rare pairing, *Endeavour* sits on the main runway at Edwards following its landing (STS-68) while *Columbia* flies by on an SCA to KSC. *Columbia* had just come out of long-term refurbishment at Plant 42 in Palmdale, CA. *Endeavour* is not yet ready to tow. NASA EC94-42789-4

[55]Smithsonian Institution's Udvar-Hazy Center at Dulles Airport in Virginia.

long after that. He really had a good time.

We did a double ferry, the ferry from hell, that's what I call it. It was a tough, tough trip. I had it better than the guys on the other airplane. I took 905. We went over to Palmdale to pick up *Columbia*; in the mean time *Endeavour* had landed out here, so they had to turn *Atlantis* around and get it ready, and we were ready at the same time so we left together for Florida. This was a double ferry. 905 was over in Palmdale and took *Columbia*, and then 911 with *Endeavour* took off from Edwards.[56] Well, they both took off at the same time but they wouldn't let us get close to each other, for security reasons. That would of made the best picture NASA would ever have had, I think.

Gelzer: Not even over the base they wouldn't let you guy's get together long enough to get a photograph? That would have been an extraordinary picture.

Seidl: Oh, it would have been. I just can't believe they wouldn't do that, but security, they were really scared back then, I guess.

Gelzer: That's when they still kept the two airplanes apart.

Seidl: Yeah, actually we started parking them together but not real close together, like one down in Area A and one up on the flight line.

Gelzer: In Marana, Arizona?

Seidl: No, here.

Gelzer: Oh, you mean down at the shuttle area and one at the back ramp.

Seidl: Right.

Gelzer: For a long time they insisted on keeping one in Arizona and one here.

Seidl: Yeah, actually they were doing that for along time and then, when the contract with Evergreen expired and they let it go, then they started parking them together. Of course, when 911 first came, I don't know whether you know this or not, but it was at El Paso, Texas, for a couple years. From '88 to '91 they flew it out of El Paso. The El Paso guys took care of it. That's where we stripped the interior out of 911. They said it was going to be over $2 million to do all the work, stripping it out and doing all this. Well, we did it ourselves. We would send people from here to El Paso to strip it; we'd take turns. One would go every month. Anyway we stripped it out, it probably was about two months or so.

Gelzer: To strip the airplane?

Seidl: Yeah, to strip the airplane. It was kind of fun stripping it because you didn't have to put it back together.

Gelzer: So you don't care. The crew that did this came out of the Dryden shuttle area?

Seidl: Yeah, it was our people. We did it.

[56]*Endeavour* completed STS-68 and landed at Edwards/Dryden on October 12, 1994. Its return to the Cape coincided with the completion of *Columbia's* overhaul and return to the Cape.

Gelzer: When you were flying on the double ferry did you guys land at the same bases for fuel?

Seidl: No, no they wouldn't let us get together at all. As a matter of fact, I was in 905 and we went to Dyess AFB [Abilene, TX], and the weather was really, really bad during that time, and of course 911 ran into the same thing. They ended up going to Altus AFB [Altus, OK]. So they were in Altus and we were in Abilene. We were sitting there pretty good because we didn't have to run the purger or anything like that because the shuttle just came out of Palmdale.[57]

Gelzer: So you could actually fly on the SCA instead of the Pathfinder?

Seidl: Right! And we did, but the guys that went to Altus had to keep the purge units going all night long because it was cold weather and it was really a mess. They were having such bad weather they were expecting hail so they had to put it in a hangar. The Air Force at Altus got them a hangar--that's a big hangar. I forget what it is like--68 feet tall or something like that. Anyway they found a hangar and got them into it, but they had to keep the power on all night and it was really rough for them. When it reached 45 degrees they would have us start all these purge units and all electrical power and all this stuff to keep the orbiter at a constant temperature.

Gelzer: To keep the orbiter warm?

Seidl: To keep the orbiter warm. Yeah, right! That was really, really, really hard. After the four days were up we took off and we went to the Cape. On the way to Cape Canaveral, instead of X-68 we landed at Cape Canaveral AFB, which is real close. It's just down the road. So we flew over there and on our way, it was kind of funny, they wanted us to fly over the capital, Tallahassee, Florida, so we flew over Tallahassee and people were running out of the capital buildings and all. I thought I saw Jeb Bush [governor of Florida] waving, I think I really did. I tell you what: you could see those people. It was really funny. Anyway they were like little rats coming out of buildings.

We finally got in there, and then we had to wait for 911 to land at X-68 and off load. They went to Barksdale AFB, Louisiana, to Eglin AFB, in Florida, and then they stood by at Eglin and waited for the weather to clear and then they snuck into X-68. So they were off loading and we was waiting over at Cape Canaveral; they had to finish because there wasn't enough room for both airplanes and both orbiters. After they off loaded it, they flew the airplane over to Cape Canaveral and then we took the other airplane, flew it over and off loaded it and then they took off the next day from Cape Canaveral and we took off from X-68 back home.

Gelzer: How long does it take once you take an orbiter back to Florida to unload it from the SCA?

Seidl: Well, sometimes it takes quite awhile, but it usually takes one night or one full day before they can get it off of us. It takes quite awhile.

Gelzer: Almost as long as it takes to put it on?

---

[57]Having just come out of a full overhaul, *Columbia* was inert. *Endeavour*, however, having just returned from orbit, still had residual fuels on board. Moreover, any experiments or equipment that came back were still in the cargo bay; as such the orbiter would need fresh air pumped into the cargo hold to maintain positive pressure.

Seidl: Well other than the . . . Yeah it takes as long to get it off not counting all the prep work. The prep we're talking seven days or eight days.

Gelzer: Once you have the orbiter lifted in the MDD it can take a long time to drop it down . . .

Seidl: Yeah, it takes quite awhile but the main thing is getting all the bolts out and getting everything ready; the bolts that hold it on.

Gelzer: There aren't that many of those. Why does it take so long?

Seidl: It's just safety. It just takes a long time. It's amazing to me too, but it takes a long time.

Gelzer: And then lifting it of course is not going to happen in a hurry.

Seidl: Yep. And then they have to do the hydraulics and let the wheels down and everything after they get it up. Of course, then we have to move out from under it after that and they put all that down. It takes them awhile. They're slow.

Gelzer: Tell me about *Atlantis* in '01 here in the MDD.

Seidl: Oh, that was actually the trip. That was before they took off from here. The next one was *Atlantis* again, taking it back after another mission. The weather en route was bad so we ended up in Amarillo, Texas, first, and then Offutt AFB, Nebraska. I think there was a whole bunch of weather south so we decided to go up north and go around it and when we went up north we ended up being in Amarillo, Texas, then Offutt, and then Fort Campbell Kentucky. Fort Campbell was kind of interesting. This was the 3rd of July so it was almost the 4th of July, and they let all the schools out and they had hundreds of school buses with all these kids there to see the shuttle. It was just really amazing to see. They had so many kids we were giving tours to. I just couldn't believe. They loved it.

Gelzer: When the SCA is in flight and it's got an orbiter on the back does it deviate for traffic or do the air controllers move traffic out of the way?

Seidl: No, they move traffic pretty much out of the way. We're so low we are usually under traffic. The only thing we have to worry about are news helicopters. When you go in some place they're flying all over the place usually. That's the main concern. Those people were real appreciative too.

Gelzer: What do you remember of the return trip from here to Florida, in which I think Gordon Fullerton was flying the 747, when they had to come back shortly after taking off because they had a light, an engine fire light?

Seidl: They had an engine fire light, yeah. Let me see. I remember that happened . . . as a matter of fact we had to change engines real quick.

Gelzer: You had a fire then?

Seidl: Yeah, we had a fire. We took one engine off, this was really amazing: we took one off of 911 and put it on 905. The only thing is, when you take one off of 911 it has things on it that 905 doesn't have so we had to do a mod to our airplane, a service bulletin modification to rewire things so we could do that. We did it and turned it around in about three days or so.

Gelzer: Were you able to explain the source of the fire?

Seidl: Actually they did. They said it was a stuck valve and all that hot bleed air that was blowing past started the fire on the wires in the compartment.

Gelzer: It was an electrical fire then?

Seidl: Yeah, yeah and it burnt up. It burned quite a bit of wires and stuff. Luckily it was all stuff on the engine so . . .

Gelzer:. Did it land on the lakebed?

Seidl: No they went back to the runway. They don't land on the lakebed very often with that 74.

Gelzer: I'm just thinking in terms of a heavy airplane.

Seidl: Actually, now that you brought it up, we had to do a max gross weight inspection because he was over the limit because they didn't have time to dump all the fuel.

Gelzer: What's the strangest thing you can remember about the SCAs? Let me back up this way: you've got two airplanes. Airplanes tend to have their own characteristics, even though they are made by the same company and have the same certification basis, they are not the same: they have their own characters.

Seidl: As a matter of fact 911 has the dash 200 gears and they're bigger gears and the stopping was amazingly better on 911 then on 905. You could really tell the difference once they landed: you could stop a lot shorter with 911 than 905.

Gelzer: Were there other peculiarities with the airplanes; one was easier for this and one was easier for that?

Seidl: No, I don't think so. You know, I never heard the pilots say anything

Gelzer: Or the mechanics complain about difficulty dealing with one and not the other?

Seidl: No, they're pretty similar. They're very similar. Of course I had a preference to 905 because it was the first and had low hours. It had a lot lower hours on the engines. It was a lot better airplane, I thought.

Gelzer: Why did you think it was a better airplane, apart from the fact it was the first and that's the one you started with?

Seidl: Actually, it was a lot cleaner and didn't have so much corrosion. The Japan Air Lines (JAL) plane had a lot of corrosion hidden in different places. As a matter of fact, after they'd bought it--Skip Guidry picked these airplanes out--he's kind of the flight engineer plus he's the maintenance chief on the airplanes, he kind of went and looked at them all and chose the airplanes that NASA wanted to get. Well, he thought

The end of the emergency exit tube in 905 that ran from the upper deck to the cargo hold. It was installed to give the SCA crew an escape route should something drastic happen with *Enterprise* on board; the crew was to slide down the tube and out of the plane; once clear, they could open their parachutes. NASA E76-29902

he was going to get fired because 911, after they got it back, started looking like they had dished out around these rivets because of corrosion. We took the corrosion off and it was all on top of the stabilizers, the horizontal stabilizers. Anyway, he thought they were going to fire him over that because they had to change the skins on the 911.

Gelzer: The horizontal skins on the 911? They were going to put the end plates on anyway.

Seidl: Yeah, so it all ended up okay, but he told me that story.

Gelzer: 905 had the tunnel that went from the flight deck to the main deck and then down out the bottom for the ALT.

Seidl: Right. They took most of that out. It was mostly all taken out.[58]

[58]While preparing for the Approach and Landing Tests there was concern at Dryden about the possibility that *Enterprise* might, on separation, slide back and strike the vertical stabilizer of the 747 instead of gaining altitude and flying away from the mothership. Dryden insisted on safety steps and modifications to the 747. The flight crew wore parachutes. Most remarkable, a hole was cut in the floor of the top deck behind the cockpit and a corresponding hole was cut into the main deck of the airplane almost directly beneath the first hole. A segment was prepped in the side of the 747's fuselage so that a panel could be blown out. Finally, a tube ran from the upper deck to the cargo hold where the side of the fuselage had been prepped: in an emergency the crew was expected to slide down the tube and out the hole in the side of the airplane.

Standing on the main deck of the 747 Dryden center director Ike Gillam looks distractedly at the ceiling while two other VIPs chat. Behind them the emergency escape tube runs from the upper deck, through the floor, down to the side of the aircraft in the cargo hold. NASA E77-31893

Gelzer: You can still see the grate in the floor.

Seidl: Right, you could see the plate put over to block the hole. Both airplanes are really outstanding airplanes. They're both good maintenance-wise. We had a lot of Inertial Navigation Unit problems with 905 at first, but then they upgraded and went to the Litton 51s and then they went to the 92s; after they went to the 92s most all our problems went away.[59]

Gelzer: They acquired their first 747 in what year?

Seidl: '74.

Gelzer: This is only two years after they entered commercial service.

Seidl: Yeah, they got it at a real good deal. I think they bought it for $16M or something like that, something ridiculous, and the reason was, the gas crisis at that time and

[59]Litton Aero Products LTN-51 Inertial Navigation Systems.

they put it in storage. That airplane was in storage when they sold it, when American Airlines sold it.

Gelzer: So American Airlines was willing to separate from the airplane because of the price of fuel?

Seidl: Yeah. So NASA got it for a song and a dance. They really did well on that, and then of course 911 came from JAL. It had a lot more hours and a lot more cycles on it than 905 had.

Gelzer: But that didn't dissuade Skip from picking the airplane and that wasn't going to upset anybody in terms of the cycles. They didn't seem concerned about the cycles. It's like the NB-52B: there weren't going to be many cycles after that were there?

Seidl: That's right!

Gelzer: Not compared to a commercial airliner, anyway.

Seidl: Right, it ended up being a good buy. Like I said, both of them really operated well and did the job. That was the main thing.

Gelzer: Why is the SCA with the shuttle on back not allowed to land at night?

Seidl: That was just one of the rules that they came up with, no night landings, no visible moisture. I guess that's one of the reasons--you can't see visible moisture at night. The one time you probably heard of it, you might not have, Joe Algranti was flying and they had the G-1 from JSC as the Pathfinder and it's slower.[60] It's a few knots, like five knots, slower then what we are. We're at 250 knots or .6 Mach and it was just a little bit slower and we took off probably about 15 minutes in front of the SCA because the weather was bad. We decided, 'well we'd get off and be a Pathfinder.' Well, is it wasn't long and here comes our airplane by us. It got in front of the Pathfinder and there's nobody to look for weather for them. So they ended up going through a thunderstorm costing a million dollars worth of damage on tiles, just about.

Gelzer: What were they carrying? Which one were they carrying?

Seidl: You know, I forget which one, but they ended up with a lot of tile damage. It might have been *Columbia*. I'm not real sure, but they damaged a bunch, 'cause that rain of course takes that leading edge stuff and just beats it away. It's funny how they can come through all the temperature and come back to Earth and yet a little rain will just peel that off.

Gelzer: Those raindrops are hitting that at 250 MPH; they're going to hurt. They must have been flogging that G-1.

Seidl: I tell you what: if it had been anyone else other than Algranti there would have been lots of trouble. He had a way of getting out of trouble.

Gelzer: He goes back to the Lunar Landing Research days in the 60s and the accidents they were having at JSC

[60]The G-I, or Gulfstream I, was designed as a twin-engine corporate turboprop aircraft.

Seidl: Yes, as a matter of fact he popped out of one of those Landers.

Gelzer: Yeah, he and Bud Ream were among the ones that had to eject.

Seidl: Yeah, yeah, yeah…

Gelzer: Your comment about keeping schedules makes me wonder if that wasn't a factor in retrospect.

Seidl: It very well could have been. I wouldn't be surprised. Joe had his way and if you know Joe, it was his way or the highway.

Gelzer: Tell me about the limitations on the SCA when it's carrying an orbiter.

Seidl: It's 250 knots and the .6 Mach.

Gelzer: Is that enforced because of the altitude you're flying at or is that because of the orbiter?

Seidl: Actually I don't know. I think it's the airplane, because with or without the orbiter we are at 250 knots.

Gelzer: Regardless of altitude?

Seidl: Yes.

Gelzer: Oh, because of those struts and those end plates.

Seidl: Fins… The fins, now that's the weak point on the aircraft, back on the horizontal stabilizers: about ¾ of the way out, there's a big splice plate put in there and that supposedly is the weak link of everything. So we used to eddy current every time at Evergreen, you know we went into Evergreen for maintenance, they would eddy current around those fasteners in that strip plate. I'm sure they probably still do. We used to do that regularly just because of that. I was going to say in the crosswinds, I forgot what is was, I was going to say 25 knots crosswind max. It didn't have very much crosswind capability. It couldn't take very much.[61]

Gelzer: When they first took *Columbia* back, were you here for the first trip to Florida, for its prep? I'm curious about the number of tiles that were missing by the time you guys got to Florida.

Seidl: You know, I don't know if there was any missing or not.

Gelzer: They had trouble initially with sticking the tiles and making them stay.

Seidl: Initially they did. As a matter of fact there might have been 3 or 4 tiles that came off but I think after that they really got them down because they changed their room temperature vulcanizing (RTV) stuff. Later flights we didn't have any problems at all. But I think you're right there was a problem at first.

[61]Evergreen International Aviation provided aircraft maintenance services to NASA for its Boeing 747s in Marana, AZ, where the firm has a large presence.

Gelzer: Was that early flight the one you were talking about with Joe Algranti in the G-1? Was that the only time they went through moisture or do you know of other trips that they made?

Seidl: That's the only time they went through rain. I can't think of any other time we've even came close. Sometimes when we come in to X-68 you could see clouds around but they were really easy to avoid.

 Here was an interesting tidbit. I think it was 905; we were out on a local flight and it was getting pretty close to a mission time. The pilots were getting their proficiency time. Well, we hit a buzzard with the number 3 engine and it ruined 13 fan blades. The pilots went ahead and flew the whole practice missions with . . . we probably had about 12 landings or so; flew and flew and flew and came back and then we saw this damage.

Gelzer: In other words the engine kept running.

Seidl: The engine kept running. It wasn't out of balance or anything. It was just amazing and these blades were just jagged. I could not believe it when we got home and on the ground I went and looked at that engine and saw all those jagged blades.

Gelzer: Did you know you hit the buzzard or was it only afterwards?

Seidl: No, it was afterward. After we got down is when we knew. It was just unreal. It was amazing. We did an inspection on the engine core. I guess when you hit a buzzard with those big fans it throws everything out, but nothing went through the core engine. They were planning on going to Pratt & Whitney and getting a new engine, having them change everything, but after we borescoped it and found out there was nothing wrong with the core, we just ordered a set of blades, which was very expensive anyway. We had to order the whole set of 36 because of the balance, the weight and balance had to be perfect on them so you order them by sets. So we changed the whole set out and we were back in action about, it was about 3 or 4 days.

Gelzer: In between uses of the SCA, especially once they moved the 747s back to Dryden and they kept them here and they performed the maintenance on them here, what went in to keeping the airplanes flight worthy?

Seidl: Uh! The main thing was to keep up service bulletins, Airworthiness Directive notes, and at the end it ended up being a lot of corrosion work packages.

Gelzer: Even though the airplanes were in the desert?

Seidl: Yeah! We found a little bit of corrosion but nothing like an airliner, especially like JAL would have. Yeah, they decided to do these corrosion packages and they were really a lot of intensive maintenance. Tearing stuff apart, looking at it and putting it back together.

Gelzer: You have a note here about the temperature in the summer and having to go inside the wing cell.

Seidl: Oh yeah! We had a leak one time on the right wing of 905 and had to get inside of it. We were out there and I just… you could not touch that wing without burning your hand. Anyway, we had to get in there and do maintenance on it. So we ordered some big air conditioners and finally got it cool enough inside that you could get in it, but it was really tough. We had to open that whole wing up and purge it really well. That was just a nightmare, but we got it done without anybody falling down in the tank.

Gelzer: I'm surprised you guys didn't work at night.

Seidl: Actually we did change. We had night people but we also had day people.

Gelzer: Oh, so it was around the clock?

Seidl: Yeah, it was hurry-up-and-get-it-fixed-so-we-can-haul-our-orbiter. Evidently the orbiters were going so often that every time you broke an airplane you had to get it fixed quick because of impending ferry coming up.

Gelzer: After the *Challenger* accident in '86 it took awhile before they flew again: what did you do with the airplane during that time?

Seidl: Actually it didn't affect us too much. The pilots still had to keep up their proficiency so we would still fly once a month so we kept pretty busy during that time.

Gelzer: So your life really didn't change because you had to be ready to ferry?

Seidl: Actually we had more time for maintenance and it was probably better on us; we could get a lot of maintenance done instead of all these trips.

Gelzer: Which comes first, age of the tire before it gets replaced or the tire wears out?

Seidl: Oh, the tire wears out. There's no aging on this. Those tires don't last that long. The tires go pretty quick. We have been real lucky with tires. I think they had a CV-990 that burned up on the runway one time that belonged to NASA. It caught fire because of brake, heated brake and stuff and the tire blew and it went up into a wing.[62] After that--now this is JSC--they decided that you could not intermix tires. They all had to be one brand name, like Michelins or Goodyear. You had to have all one brand and no recaps. For a long time, when we first came on, we were doing recaps and we did throw a couple recaps. It was usually minor damage. It was like to a gear door or something, maybe a dent.

Gelzer: How many landings do you get with the orbiter on back?

Seidl: You know, the tires go pretty quickly, especially the body gear, the inboard gear, and the first to touch down. They go pretty quickly. You probably get maybe 26 - 30 landings on them and then they're gone. You don't have to worry about a tire getting old on that. They don't weather.

Gelzer: What wears out--if anything--faster than you'd expect?

---

[62]Two tires on the right main gear of NASA CV-990 (N712NA) blew during take off from March AFB, CA, at which point the crew aborted the take off. During deceleration debris, either from the tires or from the wheel and brake assembly, was thrown up and penetrated the underside of the right wing, puncturing the right wing fuel tank. "Leaking fuel ignited while the aircraft was rolling and fire engulfed the right wing and fuselage." The aircraft was a total loss. Accident Aircraft Report: NASA 712, Convair 990, N712NA, March Air Force Base, July 17, 1985, Executive Summary, (NASA, Washington, D.C. 1986), 2.

Seidl: Well the paint jobs, being out here in the desert, they don't last very long. I'm really amazed how they deteriorate pretty quickly and you get the oxidation on them. But NASA usually paints them. They have a schedule and they keep them looking pretty nice. The paint jobs go pretty quick out here just because of the hot weather.

Gelzer: You were here when they actually went from plain aluminum to painted airplanes?

Seidl: Yes. Uh huh.

Gelzer: And the mission markings disappeared . . .

Seidl: Oh yeah.

Gelzer: And they never came back. Any idea why?

Seidl: No. I think Dryden was kind of the pusher of the missions; JSC, I just don't think they wanted them. They wanted a clean airplane, so they just decided not to put mission markings on them, plus doing that would be a lot of . . . after awhile it would be a lot of missions.

Gelzer: You'd do a lot of stenciling on the side . . .

Seidl: After about two or three hundred missions it probably wouldn't look so nice. Yeah, they just never did it. But NASA keeps their airplanes looking pretty nice. They do a good job.[63]

Gelzer: Anything humorous that comes to mind regarding the airplanes and ferry flights?

Seidl: I think the most humorous thing I guess was to watch the pilots. It was kind of funny to watch Algranti and Fitz Fulton--they both were bosses. Fitz out here at Dryden and Joe back at JSC, they kind of had a lot of clashes and I kind of enjoyed watching them argue and fight it out. Matter of fact, Paris was one of them. Joe was supposed to go home after he got it there. His daughter was graduating from school and he chose to stay. Of course he was in trouble with his wife on that one. Him and Fitz got nose to nose on that one.

Gelzer: He wasn't flying the airplane; he was just overseeing the entire trip right?

Seidl: Yep.

Gelzer: Who was flying the 747 for the Europe trip?

Seidl: Actually they took turns.

Gelzer: They did?

Seidl: Yeah and they flew people over there and swapped out pilots. I had a couple trips

---

[63]In mid 2012 Dryden reapplied mission markings to 905 in a condensed form, in much the same way as they were on the NB-52B 008. 911 had already been retired from service.

myself over the years. I went to Sydney, Australia. Qantas redid our engines. Every time they'd get ready to run an engine they would need somebody to come over and verify that it would run okay. I got a couple of those trips to Australia. Sydney was really nice.

Gelzer: It was cheaper to have the engine redone there then it was to have them done somewhere in the U.S.?

Seidl: Yeah, amazing isn't it? Pratt & Whitney over at Hartford, CT, they wanted so much money that we went out for bids and Qantas was cheaper, so that's where they went. And what was really nice about it is Qantas has all these freighters coming back and forth to L.A. so they would take our engines and ship it free. They would just put it in. They had what you call that fifth engine pod and they would just put it on and fly it back and some of them would even put it inside the cargo bay if they had room.[64]

Gelzer: Yeah, but it's much more interesting if you're moving around the sky with a fifth engine pod.

Seidl: I think they might of done one that way. Just room I guess. They had enough room in the cargo bay they would use the cargo bay. That was fun. I had a trip to Montreal, Canada, for the same reason, to cover engine runs. That was a good time there. During the one in Sydney, one of the guys was a speedboat enthusiast so he took me out to watch the speedboats. I didn't know it at the time but I ended up in the in field on a yacht and all these speedboats going around. What a life!! I had a good time.

Gelzer: What did [Richard] Dick Scobee do?

Seidl: Dick Scobee was our pilot on the 747.

Gelzer: Really?

Seidl: Yeah, he was.

Gelzer: We're talking about the same Scobee that was an astronaut?

Seidl: Yeah, uh huh. He was our pilot, even in Paris. June Scobee (his wife) was in Paris with us. That was funny. We were in Paris and this driver was taking us and of course we had one woman and about six guys in a van and he was driving us around and showing us around. Well, he went down to the Place Pigalle. I don't know whether you know what Place Pigalle is, but it's where all the prostitutes line up on the street. Anyway, we had June with us and we were going along and June said: "this is no fair." I asked: "what do you mean, no fair?" and she said: "well, where are the men?" (Laughing) She was quite gal. She was very nice.

Gelzer: Actually, Dick Scobee flying is not that different from [Gordon] "Gordo" Fullerton who flew heavies—he was actually flying the parabolic flights for a while and then flew the shuttle, so it's the same thing.

Seidl: Yeah. Dick was really good. You know, they were having tire problems with the

---

[64]Airlines have often moved spare jet engines from one location to another in what is dubbed a "fifth engine pod," which is a pod enclosing the engine that is typically hung inboard of engine no. 2 on a four-engine airliner.

grooved runway out at the Cape; he was the only astronaut I know would come out with his own camera and take pictures of our tires after we landed on it and then compare it with the shuttle tires.

Gelzer: Tell me what the differences were? How bad did the tires get torn up on the original X-68 runway?

Seidl: You know, our tires (on the 747) were okay and I don't know whether it was because the air pressure or what. It would scuff them where they first hit but it wouldn't tear up the tires like it was doing on the orbiters. The orbiters were probably a lot harder, I think.

Gelzer: Between 340 and 400 psi.

Seidl: Well, that's uh. 400 . . . ours were usually at 320 I think.

Gelzer: So it's not that far off.

A. Yeah, except that 911 had bigger balloon tires and those were a lot better tires. They wouldn't wear quite as bad as 905s.

Gelzer: And they're more tires to displace the weight as opposed to just four main on the orbiter.

Seidl: Yeah, right and that probably had a lot to do with it.

Gelzer: So Dick Scobee would go out and take pictures with his own camera of the tires to compare?

Seidl: Yeah.

Gelzer: Of the orbiter and 747.

Videographer (Parcel): Gordo right?

Seidl: No, it was Scobee. Dick was quite the guy. Of course I don't know if you know, he came up Air Force. He was enlisted in the Air Force. He was just a mechanic and went to Officer Candidate School and got to be an officer and started flying and went right up the ranks and ended up being an astronaut. He came a long way. It was really too bad about that.[65] I kind of wonder sometimes about the thought behind launching on a cold day like that. You could see the icicles on TV. Of course, anybody can second-guess.

My wife, just about every flight we had, she'd bake something. She had treats for us. She made the best brownies of anybody. In all these years I would hate to guess how many brownie pans she went through. It was amazing.

Gelzer: Baked for the crew of the SCA and the other people?

Seidl: The SCA crew, just the SCA. I would put them on the SCA and everybody would

---

[65]Francis R. "Dick" Scobee was shuttle commander of 51L, *Challenger*, which exploded barely more than a minute after launch, January 28, 1986.

get them after flight, the pilots would eat it during flights. Every mission, she never missed. I think I would get a list and see how much brownies cost. Maybe put in a bill for that.

Gelzer: Who you going to send it to? I think it's a good idea.

Seidl: I don't know either.

Gelzer: Miscellaneous expenses.

Seidl: Yeah, yeah.

William T. "Bill" Shelton and Harry Talbot

Harry Talbot retired from the United States Air Force Reserve with the rank of Colonel following 31 years of active and reserve service. His last assignment was as the Mobilization Assistant to the Commander at the Air Force Flight Test Center at Edwards Air Force Base in California where he was responsible for advising the Center Commander on issues related to the utilization of reserve forces, Space Shuttle recovery, emergency response management environmental issues and Budget. His career also included assignments in Information Management, Transportation, Security Police and Air Base Wing. Talbot received his B.A. degree from San Jose State University in California, his M.A. degree from Claremont Graduate School and completed the Air War College.

William T. "Bill" Shelton worked as a Range Control Officer and was the only civilian at Edwards Air Force Base dual certified as the Space Shuttle On-Scene Commander and for control room duty. An FAA certified air traffic controller, Shelton worked for many years at Edwards, where the agency maintains an air traffic control station that controls the air space over the base (SPORT). Shelton earned a B.S. from the University of Redlands and an M.S. from the University of California in Systems Management and moved into plans and policies as the shuttle program wound down. At the time of this writing he a Senior Technical Advisor in the private aerospace industry.

Bill Shelton and Harry Talbot, interviewed by Christian Gelzer, June 7, 2011, NASA Dryden Flight Research Center, Shuttle History Documentary Interviews.

Christian Gelzer: It's June 7th 2011 and I'm here with--

Shelton: Bill Shelton

Talbot: and Harry Talbot

Gelzer: We are here to talk about shuttle and the Air Force's role in that program. I'll allow each of you to introduce yourselves in terms of who you are and your background.

Shelton: My name is Bill Shelton. I showed up in Edwards in 1979 as an FAA controller. I was probably the only person working on Edwards Air Force Base that did not see the shuttle land out on the lakebed because I was working for the FAA at the time. I was the air traffic controller and had all the air space; I was trying to keep people away from the shuttle because everybody wanted to come in. We probably had 20 or 30 airplanes over by Palmdale circling just wanting to see the shuttle come in and land, so I was trying to keep everybody out of the air space. When it came in everybody ran out on the roof of the building except me, so I didn't get the chance to see the shuttle.

    I left the FAA for a couple of years, came back in, and worked over in SPORT, our military radar unit or Air Traffic Control facility [at Edwards]. I eventually [became] the deputy of the operations support squadron, and then, from there, as the person who was kind of guiding base operations and airfield management and air traffic control and that sort of thing with my background as an air traffic controller.

Part of our duties was to work with the shuttle program as a contingency coordinator with the different entities that came in to support shuttle landing or an exercise, and that just sort of grew into helping out and operating in the control room with John Evans, who was kind of our predecessor in some of this. Eventually, through transitions of colonels coming and colonels going the decision was made that a civilian would be a good idea [handling the shuttle activities for the Air Force] because the civilians are here; it's not like we rotate through every two years. I got to be the on-scene commander, working between 10 or 12 landings, something like, that here at Edwards.

Gelzer: You were always a civilian in that capacity?

Shelton: Yes.

Talbot: I'm Harry Talbot. I came here in November of '81 as a brand new 1st lieutenant out of the National Guard. I'd been looking for a different reserve assignment, came over here and my first day of duty at Edwards was as a lieutenant in crowd control on the East Side [of Rogers Dry Lake], the "East Side Dust Suckers." Envision that side of Edwards Air Force Base with 500,000 people, more RVs than an average city has, similar to the Wild West. I stayed on the East Side, stayed working the East Side through Return To Flight [STS-26B, September 29, 1988, *Discovery*].

After Return To Flight, after *Challenger* crashed, by this point I was a captain, and I moved over doing close end security on the orbiter with the security forces. Every two years you got a new colonel, and by that point the shuttle, I don't want to say was losing some of its luster, but spending an entire day working shuttle was hard on a colonel's schedule. They started out with the Air Base Wing commander in [the Convoy Commander's Vehicle] then it went down to the Air Base Wing vice-commander. Someplace along the line, when I was running the close end security, they said: "hey kid, come here, we're going to have you drive my car out here," and then: "I'm going

An aerial view of the crowds on the eastern shore of Rogers Dry Lake that came to see the first landing of the shuttle *Columbia*, April 14, 1981. The area opened to the public was very near the site first used by the military on a seasonal basis in the 1930s and designated Muroc Bombing and Gunnery Range. NASA ECN-18978

to teach you how to do tow," [back to Dryden] so I did that for a number of years. Then I left the security forces and went over to the Air Base Wing as the reservist to the Vice-Commander. As part of that they checked me out doing shuttle, being in the command position, and much like Bill, I had been doing it for awhile. The active duty [officer] would come and the active duty [officer] would go and finally one of the new Deputy Air Base Wing commanders said: "look, you've been here for a long time, you know what's going on. This is an ideal job for you because we just come and go." So I took over the whole operation in about '94; by that point I was a colonel. And then, in '99 when I was offered the job as the Individual Mobilization Augmenter [IMA] to the commander, the senior reservist on base, I said I'll take it but only if [they'd] let me keep shuttle; so I kept shuttle, Bill and I. Bill had been running the operations side for years; I had been running the close end support: fire, police, and medical side, for years. We joined up as a team and spent the last four years – five years– doing shuttle together.

Gelzer: I want to get some clarification before I come back to your first day of service out here: "operations" constitutes what in this instance?

Shelton: In this instance we're referring to Operations Support Squadron [OSS]. It's internal to Ops Groups, which is internal to the Test Wing, and it takes care of airfield management, the control tower, SPORT. We do the weather observing, we do life support, we do center scheduling, and we do everything cradle to grave, if you'll excuse the expression. You want to go fly, you have to go scheduling, which is an OSS function. If you want to go ahead and start to taxi, you call the tower for taxi. You know, when you go fly out in the air space you are still talking to SPORT, and so we tried to get our hands around everything that was operations oriented, put it in one organization so [that] if there were any coordination issues it would be easy to resolve inside one organization rather than hodge-podge out all over the base. The only thing we didn't own was the [test] range itself, and that was over in the Range Squadron or the Range Division or whatever it happened to be.

Gelzer: Your first day here you were basically sent over to East Shore?

Talbot: Yeah, the East Side Dust Suckers. It was before the new security building, which is now the old security police building, was built. We had these trailers: I showed up for work, it had to be 1 – 2 o'clock in the morning, I wasn't checked out with a weapon, I barely knew where anything was; they gave me a badge and a beret and I was assigned to a staff sergeant. We got in the jeep-or squad car I think at the time-and we went over the Santa Fe Trail. The Santa Fe Trail is this dirt trail that crosses the lakebed.[68] Well, I didn't get over there until about 3 or 4 in the morning. [It was] pitch dark, people everywhere and we were wandering around because we were definitely out numbered, I mean we had maybe, maybe 50 to 60 cops, not more than that.

We had some National Guard guys in jeeps, which were really scary. We had radios that lasted 6 to 8 hours. By that point we were operating out of a tent. There was a small fence around the place and it was just such a festive atmosphere. There were vendors; there were rows of the famous Edwards porta-potties held together with ropes because it's always windy, and row upon row upon row of 40 – 50 foot long RVs. I

---

[68]The trail is actually the remnant of the Atchison, Topeka and Santa Fe railroad that crossed Rogers Dry Lake, coming "ashore" at the town of Muroc, near where the current Edwards AFB control tower stands. The line was relocated in 1953, providing the Air Force with full use of the lakebed.

saw more dollars worth of RVs than the shuttle was worth at some point. You had a fire truck or two, you had a medical area fenced off and we had the cops and there was this incredible sense of camaraderie. The public loved us and with that particular landing I don't remember any serious incidents, but it was a blur. You got out there and you had dust and sand everywhere, your ears, and your eyes. By the time the day ended you had dead radios, youngster's driving jeeps–scary–you had cops just trying to do the best they could. Everybody was just working together. We knew, the security forces people, knew what we had to do. The commander was out there, everybody was out there but nobody could talk, so what you did was, you found a problem that needed fixing and you fixed it. Whether it was the young ladies sunbathing on top of their vehicles, (who were getting a lot of police support at the time), or whatever it was, you fixed it.

You had two roads coming in at the time, two roads going out, you had the public that had been out there for days, hot, tired; at some point we had a number of people who had been out there for days drinking. I remember at the end of that day, over in the medical area we had 3 or 4--5 cars that couldn't move and you had about two people who were just coming back to them after drinking far too much--in the sun for days--who's friends had left them; just left them. So the medical guys were treating heat injuries, dust injuries, everything you could think of. The fire people were out there helping them in any way they could and the cops were trying to keep order. It was a great crowd and I remember standing out in front of the line when the orbiter touched down and this feeling of pride to be an American was just intense. The flags were every place. It was wonderful. And then you just have this stampede of people leaving, and of course it didn't matter about roads; I'm not going to use no stinking roads! They were going over the desert, they were going all over [the place].

One of those early landings, Mercury Blvd. was being resurfaced or torn up, and there were huge potholes. We had cars that had ripped out both of their tires on both sides; it was looking like a [demolition] derby because somebody who was resurfac-ing the road didn't say: "hey, we're going to have a shuttle landing; we're going to have 500,000 people coming down this road." In the early days we had a tow truck out there; invariably people left there air conditioning on, people left their lights on, we had disabled cars, so you knew what had to be done-nobody told you what to do, you did it. We had these 19-year-old airmen making the right decisions. We were in a sense a city of . . . I think about 550,000 people out there.

Gelzer: Joe D'Agostino remembers at least one trailer, maybe two trailers that were open for business.

Talbot: Well, you know.

Shelton: It was a city with all the accouterments.

Talbot: The citizen's report read: "and the security policeman left the motorhome disheveled, pulling on his gun belt." (Laughter) Yes, they were working the crowd; they were in the motorhomes from Vegas, the young ladies topless on top of their RV [with] lots of police protection.

At one point we had a young man who decided he was going to go up in the rocks over in the East Side area and look at the crowd through the scope of his rifle. He didn't hear the security helicopter that we parked a little behind him; I wasn't there but the story was, and sometimes they're embellished, that he felt something on his ear and he pushed it and turned around and looked right down the barrel of an M-16. By the time we got him in to the squad car he was grateful not have that man with the gun next to him. I don't know what really happened, but I know that man was very,

very grateful not to have these people with guns aimed at him.

Gelzer: You had helicopter support in terms of, was it the National Guard or people from China Lake that were flying a helo or two around the crowd?

Shelton: Well, we would get support from a number of places. We would get different kinds of support. We would have SAVES, the medical transports; you'd get those from Fort Irwin. We would get just the regular transport helicopters for moving people around, doing that sort of thing. We had one come over from Vandenberg because it had an infrared sight so we could use it at night. We had the folks from China Lake come out and do video, that was part of the contingency coordinator, to make sure we had all these different resources show up at the right time for exercises and for actual landings, and stand by for landings and just to make sure that we had the right crews in the right spot at the right time. That was kind of a troubling thing for the contingency coordinator because, in a lot of cases you have the [military] people with this constant rotation and so you want the photographer, say, the one from China Lake with the camera on board, you want it at a certain spot because you know when the shuttle lands it's going to do this or that, or people are going to approach from this or that direction, so you know where you want the video placed; well, now you have to go through this whole training scenario if you've got somebody new. It was the same thing with the transports and the SAVES; no matter how many times we practiced, every mission we always had somebody new, and it was just through saving grace that we had enough experienced people that it didn't really become an issue.

That was one of the reasons we continued to push for exercise after exercise after exercise. We would train for an incident on the runway. We would train for an incident that landed offsite, shall we say, and we started getting Los Angeles County sheriffs and fire involved, because at some point if that happened we would have to get them involved. So we tried to sit back and imagine: what are the things that could go wrong--hope they don't--but what are the things that could go wrong, and if something did go wrong have we practiced it before? It was one of those issues of constant practice, constant reworking the check lists to make sure that we had everything right, that we could talk to everybody--no big issue, we just wanted to talk to everybody, to get everybody on the same frequency at the same time. Then we find out we had too many people on the same frequency; one of the things that was a constant challenge for us was trying to figure out how many people are too many on the right frequency.

Talbot: In the early days we sometimes had two dedicated security helicopters. We had four youngsters, four twenty-something's armed to the teeth flying around--you know, the runway isn't fenced, the lakebed is not fenced, so we had people coming on to it. The first landing we had the infamous red Porsche that busted through and went aiming out to the orbiter, didn't realize what would happen when a helicopter came down on you that way. He tried for years to collect money to get his Porsche redone; never did.

For my second landing we had the motorcycle tearing out; motorcycles aren't as fast as squad cars in that situation so we got him before the helicopter got him. A couple of times I had the opportunity to fly security and nothing is scary as a lieutenant with an M-16. When [President] Reagan was here [July 1984] it was horrendous. We had all these issues and the aircraft were a key part of my life at that point; that was why we could get cops in the right place at the right time.

Gelzer: Describe or explain how the responsibilities broke out, NASA--Air Force, and who was responsible for what, at what point. The orbiter is coming in, you're going to form the convoy that is heading out: at what point does either one of you or does somebody from the Air Force join NASA 25, the communications van, the lead van?

Talbot: That is what I did initially at the old fire station, Fire Station number 1. When we first started doing it we would get our team together in the Fire Station and talk about our limiting factors. As soon as NASA 25 pulled up, the On Scene Commander, "Eddie Leader" [Edwards Leader] would walk out, get in the van, greet everybody, sit down, and everyone opened their checklists. Eventually we started bringing two people there on a regular basis. We would go out to the scene.

The military was responsible--we were responsible--for security of the runway, the lakebed, for close end security; we had fire, police, medical, bio environmental, everybody there. The magic word was: "a mode has been declared." NASA was in charge until the ground operations manager uttered: "a mode has been declared" and then we took command.

Shelton: As a guy who does air space, air traffic controller, one of the things that was always an issue for us was that we tried to balance a shuttle coming in and disrupting the entire Air Force Flight Test Center mission. At the time we had only the one runway [Edwards has a second, parallel, runway now]. So you come in and land on the hard surface; you're there for 3, 4 … 5 hours and everybody understands that's important, but at the same time it does tend to disrupt the [Air Force] mission.

So what we tried to do was figure out a balance. We sometimes sat in NASA 25 or at the Fire Station; we wouldn't know that the shuttle was coming until we were actually inside the window [roughly 90 minutes notice before touchdown at Edwards]. Would it get here before our airplanes would get back from wherever they were? We're trying to do this fine balance: should we call people back, are we coming to Edwards, are we not coming to Edwards, what are we going to do, because we had requirements. We were always on this fine line.

So we would have to keep everybody apprised. "Hey, the shuttle is coming in today." You would be surprised how many people just don't pay attention to that. It really was an advantage when Harry moved over to the commander's side of the house; he had much more contact with them than we had before--now someone would actually make face-to-face contact and let them know what was happening, when it was happening and [what] the issues [were]. I would have to tell every Wing Commander: "I don't know that I can call your airplanes back in time to be here, so some of your airplanes are going to get caught out, some of them are going to have to go to Mojave, maybe some are going to have to go to Plant 42, maybe they have to land at Point Mugu if they're out over the water [at the time]."[69] The only other option is that we call our airplanes back three hours before we suspect there's going to be a landing, and we always suspect there's going to be a landing every time it's up there until an actual burn to go to Florida. It was always this "catch 22," trying to figure out when that was going to happen. We would call the airplanes in; they would come in and land; once the runway was clear we would go down and do a runway inspection and make sure there wasn't any FOD on the runway, anything that was going to do any damage to any tires

Gelzer: How many people in NASA 25 were from the Flight Test Center?

Shelton/Talbot: The two of us.

Gelzer: Just the two of you, and until you [Shelton] joined it was just one?

[69]Plant 42 is shorthand for Air Force Plant 42, a Government Owned Contractor Operated (GOCO) facility in Palmdale, CA, with manufacturing and maintenance facilities as well as runways.

136

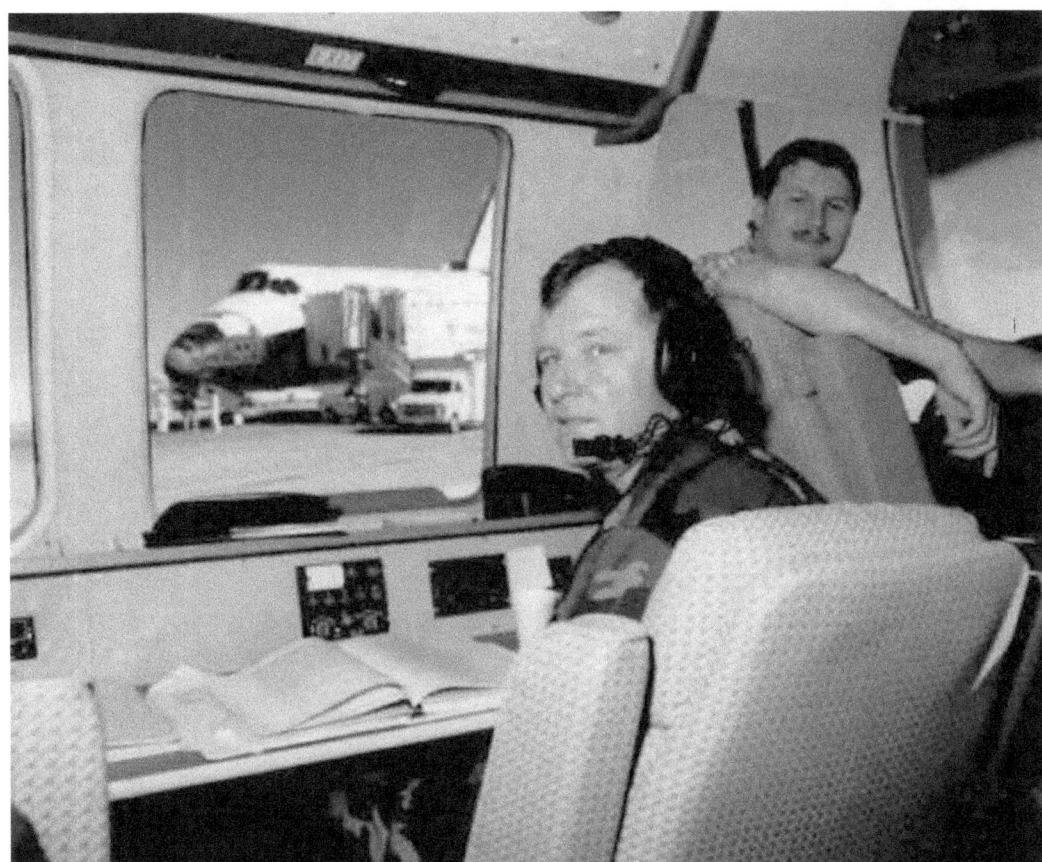

Lt. Col. Harry Talbot sitting at the console in NASA 25 and George Grimshaw in the command vehicle driver's seat. On the desk in front of Talbot is a book with check lists, steps personnel must complete en route to safing the shuttle and making it ready for tow back to Dryden. When We Were Young Photograph courtesy of Harry Talbot

Shelton: Right.

Talbot: No, Dan Reece was in there for one with me.

Shelton: Dave Sampson was in there once.

Talbot: The way the Air Force is split [at Edwards], there is an Air Base Wing, which is the mayor; fire, police, chaplain, medical, all of that stuff, and the Flying Wing, which was everything Bill did. As an Air Base Wing guy all my life I didn't care if your airplane didn't [get airborne]. It didn't make any difference to me because I wasn't responsible; I wanted to make sure that place was secure. So when Jim Roberts joined us, when we finally got an air traffic controller in there it opened our eyes to how important [it was] having both sides of the base there. We just kept going from there, and it blossomed.

Shelton: NASA is always in charge unless a mode is declared. Now, that doesn't mean we are sitting there with our arms folded. What we are doing at the same time is making sure we have the perimeter set up, that the [base] fire [department] does come in from the proper direction, that they do come in to enable any kind of assistance that may be required should a mode be declared so you don't have everybody back at the fire station, back at the security police headquarters. Everybody is there taking care of the security, fire, the helicopters; trying to keep the base up to speed on how much longer [until] the base can get back to the business of flying, that sort of thing. It was kind of a split personality; NASA was always in charge of the mission until a mode was declared, but at the same time we were positioning everybody that we thought we could get a hold of should

Inside NASA 25, looking aft. Seated at the console on the right are Billy McClure (l) and Casey Wood (r), both of United Space Alliance. Bill Shelton sits on the couch looking over procedures in his binder. NASA EC05-0079-21

something go wrong.

Talbot: The interior of NASA 25 was laid out this way: the GOM was in the passenger seat of the--the Grounds Operations Manager, Denny Gagin or whoever was there. You had a NASA radio tech who was our driver. You had one Air Force person [sitting] on the engine cover. They have a chair for us now. Bill and I were this close [shows mere inches with his hands] for hours and he was rubbing shoulders with the GOM, and right next to me was the Quality Assurance [QA] person.

Talbot: We were doing an exercise one time and I was in NASA 25--you weren't there [looking at Shelton]--and I came nose to nose with a B-1. That was the last time I took a convoy out on that runway or that taxiway without airfield management in front of me. Nothing is so real as the front of a B-1 Bomber.

Gelzer: And it wasn't parked, it was moving?

Talbot: We were both moving. I don't know how close we got. Right then and there I had a religious epiphany that I was never going to move a convoy without Larry Ledford or one of his partners in front of me.

Shelton: I think that one was because of this timing issue: he [the B-1] was trying to come back and land before [we closed the airfield].

Gelzer: What was the strangest thing you remember in connection with shuttle ops?

Shelton: Wish you'd ask me that one an hour ago . . .

Gelzer: You had something ready?

Shelton: No, but it would have given me an hour to think about it. I grew up as a kid in the 50's and 60's, space as the final frontier, all of that mystique, and here I am sitting in NASA 25 looking out the window at something that just returned from space. Maybe that's not odd, maybe that's not unusual, but for me, just a kid from Staten Island, New York, being involved in the shuttle program and looking out at a shuttle that was just several 100 yards away having just returned from space, that was awe inspiring.

Talbot: I went to Sunday school with Sally Ride; her parents were my Sunday school teachers. I can remember sitting in front of my grandparents' TV watching [John] Glenn take off; we worked his landing [STS 95, October 29-November 7, 1998].

Shelton: In retrospect I suppose some of the funniest things were the people who weren't where they were suppose to be. You would get some older gentleman who's out with his family, who takes a wrong turn somewhere and finds a hole where we didn't have a cop and finds himself somewhere where he's not suppose to be. "Hey, who's that car over there? There's not suppose to be a car over there." Then everybody jumps into action to find out what in the world is going on and it turns out to be a guy [who] flew in World War II or whatever and shows up just at the wrong spot at the wrong time and it's a good chuckle later on because of the way everybody responded and did what they were supposed to. But the look on the person's face or how the cop describes what happened when he shows up with the vehicle with the 50 caliber on the roof and the guy knows what's on the roof because he has military experience . . .

Talbot: My favorite one: were working East Shore, you know the East Side Dust Suckers, and this guy comes up in this large truck with a boat behind him and says: "Where's the lake? We want to do some boating before it lands." I said: "See the word 'dry'?"
   Then we had grandpa, it might have been after *Challenger*, with a big beautiful motorhome, he's got the wife, daughter, grandchildren; hippie in the Hillman cuts him off, scratches the motorhome. Grandpa reaches in to the glove box and gets a drop on the hippie with a 9-millimeter. We come screaming in. By the time I get there, grandpa's in handcuffs. I'm in charge at this point and I got the daughter screaming at me, I got grandpa swearing, I got the children crying, and I have to let the hippie go.

Gelzer: You're not making this up?

Talbot: No. You couldn't make this stuff up.

Gelzer: John McKay, remembers doing a check of the approach pattern before a landing and finding, I think he described it as two guys who had been--

Talbot: --oh, when they crawled across the lakebed.

Gelzer: Yeah, who had been lying under the approach pattern, and when they were challenged they said: "Well, we did this before . . ." and they'd never been caught, they had been hanging out right under the approach pattern, maybe 200 feet from the threshold of the runway as the orbiter was coming in and they had done this before.

Talbot: I got the rest of that story because they were in our holding area when I came back from the field. We booked them, tossed them inside a building so they couldn't see anything and gave them an MRE to eat; "meals ready to eat." Yeah, those guys had done it before.

Gelzer: Let's look at fire training or emergency egress. Take fire training specifically; what were you measuring your preparedness against? How did you know what they needed to do?

Shelton: There's a couple of things: one of the things that we would do on a semi regular basis with the folks out on the Cape was, we would go out with Paul Latour from the helicopter side of the house, with Joe-

Talbot: -Number two firefighter-

Shelton: Yeah, Joe was the Fire Chief. Joe was a fireman and a Fire Chief forever.

Talbot: Yeah, he use to yell at me.

Shelton: Probably for good reason, [chuckles] at any rate we would take the police, we would take the fire folks--

Talbot: --We would take the helicopter driver, we would take a medical guy--

Shelton: --So every time we went back [to the Cape] we would take some folks and we would watch how they did business back there. And they would come out and watch how we do business with an exercise or a real mission.

Gelzer: How frequently did you go through exercises?

Talbot: We did some quarterly, at least twice a year a big one.

Shelton: We did tabletop [exercises] at least quarterly and second quarter would be the actual exercise of what we walked through on the tabletop and we would do things that happened at Edwards, things that happened off Edwards: something would happen--the shuttle would land in Boron and we would practice taking everybody to the Boron airport.

Shelton: Domingo would get a chance to see some helicopters fly.[70] Maybe we're the biggest dogs in this neighborhood, but when we get outside the boundary and you get in to L.A. county and Kern county we may be looking for assistance: we need to start going to the highway patrol and the sheriff's department and fire department and stuff like that.

Talbot: Civil Air Patrol.

Shelton: All of those.

Talbot: We had a new flight surgeon and he warned me: "Harry, you probably want to be up early because I'm doing a no-notice recall tomorrow and we're going to be on the flight line next to the orbiter [a bus stood in for the real thing] at 6:00 in the morning and we're going to do this until I'm happy," and he did. We had to retrain them every couple years.

Gelzer: What was the last orbiter flight you worked?

Talbot: The crash.

Gelzer: *Columbia*?

---

[70]Domingo Gutierrez owned and operated an eponymous restaurant in Boron that was the prime destination for shuttle crews after landing at Edwards/Dryden. There is a small dirt strip airport in Boron.

The formal pre-flight photograph of shuttle astronauts John Young and Robert Crippen, crew of STS-1, *Columbia*. NASA S79-31775

Following their formal picture, astronauts Richard Truly and Joe Engle donned flight gear from another era, along with fake mustaches and leather helmets, and posed for a second photograph. NASA S80-27341

Talbot: Yes

Gelzer: Bill?

Shelton: Well, I can't remember…

Talbot: There was a landing shortly after I retired because they invited me back so you had to be working that one.

Shelton: It was a night landing and I was in the process of leaving the operations squadron and moving up to 'plans and programs' and some other folks had come in to take over, [Brig. General] Jim Hogue in particular, and I was out merely to act as a back up to the back up to the back up should something happen. I remember that there really wasn't room for me in NASA 25 because General Hogue was in there along with, it might have been Guy Dean, somebody else from the OSS. I was standing there on the flight line and I could see this star out in the West get bigger and bigger and bigger and then pass right over the base and then land; that was the first landing of shuttle that I had actually seen when I wasn't in NASA 25 or maybe [from my] back yard or something like that. That was actually my last landing--the first one I'd ever seen touch down.

Kenneth J. Szalai

Kenneth J. "Ken" Szalai began his NASA career at Dryden in 1964 after graduating the University of Wisconsin with a degree in electrical engineering. He earned an M.S. degree in mechanical engineering from the University of Southern California in 1970. Early in his NASA career, Szalai was principal investigator on the F-8 Digital Fly-By-Wire program, which successfully flew the first aircraft equipped with a digital electronic flight control system and which had no mechanical reversion capability. He also held research and systems engineering positions on several programs investigating flying qualities, integrated flight controls, and fault tolerant-flight critical systems. He became the Ames Deputy Director for Dryden in December 1990, and remained in that position until assuming the position of Director of Dryden Flight Research Center when Dryden officially resumed an independent status.

Interviewee: Ken Szalai

Interviewer: James E. Tomayko

Date: June 8, 1998

Place: Dryden Flight Research Center, Edwards, California

Tomayko: How did you start out here at Dryden?

Szalai: I started out as an electrical engineer as a new graduate from the University of Wisconsin. It was a time when there was a lot of hiring by NASA. I saw an X-15 movie, and that tilted me. I called out here and talked to Jack Fischel. He was a Deputy Division Chief, he listed all the things that were being done here. Dryden couldn't bring me out for a visit because there was no money for the government to bring out interviewees. I did get interviews with all the other places--Raytheon, Cornell Aeronautical Laboratory, Buffalo. I couldn't decide between Cornell Aero Lab and Dryden. It was very, very tough. I saw the X-15 movie that weekend. I had written a letter to Dryden; thank you but no thanks. And I was really having a hard time after watching that movie. So I dropped it [the letter declining the offer] in the mailbox; it hit the bottom of the mailbox, and I said, "I made a mistake." ....

This was a big lesson for me: You have to make decisions to find out if they're wrong. If you debate issues for too long, you never end up with anything. Actually, Mr. [James C.] Fletcher, the administrator of NASA, said, "More truth comes out of error than out of debate." That's kind of like the F-8 program. We did a lot of things. You can sit there and contemplate all of the issues for a long time, but if there is a barrier out there, somebody will be sure to tell you.

My first job was as an electrical engineer on the General Purpose Airborne Simulator, the JetStar. It was an airborne simulator. I was hired to design electrical circuits and things like that and test them and check them out in the airborne simulator.

So, in 1965, I got to spend a summer in Buffalo, and I realized that I had made the right decision - moving to the desert instead of Buffalo, which is very much like Milwaukee and a nice place. But I got interested in the aeronautical research side--not just designing the electrical systems.

One little anecdote--we had a hardware in-the-loop simulation, the first one that was set up, in a small hangar. And we had the Iron Bird that was being built up.[71] But the first simulation was just a "hot bench"--where the hardware was just sitting on a bench--not packaged for the airplane. So the AC Sparkplug people brought it in, looked around and said: "This is where you're going to do the work? There's dust, and you're going to work in here?"

"Hey, we cleaned this up before you came in," we said.

Every one of their connectors had caps on it. They said: "First of all, we're going to vacuum this whole place up." We cleaned it all up, and the next thing they did is they looked at our connectors and went [blows twice]: "These things are kind of dirty." They vacuumed everything out and checked before they would even unpack their stuff. Of course, we thought they were being a bit ridiculous. But there was the marriage, then, of space and aircraft digital flight controls. Now, really, they were right in terms of the criticality of the electrical systems. If you're going to depend on an electrical flight control, the electronics had better work. And, so, we quickly adopted their habits.

APOLLO 15 Command Pilot Al Worden and Center Director Lee Scherer stand by the display and keyboard assembly (DSKY) used in the Fly-By-Wire aircraft currently under evaluation. This particular DSKY was actually used by Worden on the Apollo 15 flight and later made available to the Center for its FBW program. Worden was here earlier this month to pick up a general aviation aircraft that was instrumented by the Center for NASA Ames.

One problem involved a restart of the computer. If you had a restart, it zeroed out all of the digital-to-analog computers for a short period, so the horizontal stabilizers went to minus five degrees for about one second. If the failure persisted, the system would transfer to the backup system. We were running some of these restart tests in the Iron Bird with Gary Krier in the cockpit and this restart occurs and the horizontal stabilizer goes to minus five degrees. The airplane starts pitching up, and Gary starts putting in inputs--but nothing is happening because in this one-second period there's no control. Gary starts to come forward on the stick pretty rapidly. Then the system goes to backup, and the airplane pitches down and he lost control. Someone told me: "This doesn't actually happen in the real Apollo computer." It turns out, it does.

We had to find a computer for Phase II. Because of the Langley interest in advanced control laws, they said: we needed floating-point arithmetic, one-microsecond multiply times, and good memory cycle time so that we could really go to town with some of these advanced control laws. But other people thought: "Let's just go out and get the simplest computer we can find that's already flying or is available" rather than these high-end computers that were just coming on the market, that had no flight experience. The IBM AP-101 was a derivative of the AP-1, which is a fixed-point computer. Singer-Kearfott had a machine, and CDC had a machine. There were three machines. The shuttle program was also going through quite a bit of analysis of what kind of machines they would want. So, they invited all the computer manufacturers to come in and bring their computers and just plug them in. NASA JSC wanted to run these computers through their paces.

I think only two companies showed up at the conference. I think the IBM AP-101 computer was the only one that actually was in a usable configuration. We were watching all of this. Cal Jarvis and I and a few other people constituted the Source Evalu-

[71]Dryden's "Iron Bird" was an F-18 minus engines that sat in one of the hangar bays of the Research Aircraft Integration Facility; it's purpose was to allow systems to be tested on an actual aircraft before being flown on an aircraft. This included the placement and sizing of new systems as well as their operations.

ation Board. The AP-101 came out ahead, so we purchased it. The shuttle program selected it sometime later. Typical of project managers, Cal was impressed that I had pre-punched holes in my notebook paper so it would fit in a three-ring binder. I said to myself: "Boy, these project managers are easy to please."

The worst problem we ever had was well into the flight program. Jim Craft calls me and says: "We failed the preflight again in the hangar." Usually, it was because of the tolerances, like the stick; you pull back, and the control surface should be 26.5 degrees plus or minus 0.3 degrees, and, sometime the rigging was off, and it was about 26.9, and the limit was 26.8. It had happened a few times. I said: "Check the rigging." He says: "We've actually checked all of that, and it's all within limits."

I went down to the hangar and started looking at the computer software from the trace

CHICKEN GIFT was presented by former DAST program manager to departing flutter engineer John Edwards. John previously made the mistake of telling Don a method of predicting flutter was to examine the liver of a chicken.

where the failure came from. We have a listing of the software that is loaded in the computer. I then start checking memory locations in the computer to look at the contents of certain locations. I couldn't believe my eyes. I looked at the memory; I looked at the listing. They were different! "The code is different! The instructions are different! We've got some instructions in the computer not the same as they are here. And they're right, in the listing. This is dramatic. Is there something else wrong in the computer?" This happened to be in a self-test routine. What if it had been in the control laws. This was an allowable time to panic.

Quickly, we contacted the Draper Lab; they said, "Impossible! This cannot happen!"[72] Turned out, it is possible. It's the assembler. This is the part of the process that generates the computer code from the original code. It creates a file that goes to the printer, for the listing. Then it takes this file and puts it somewhere else, and writes the tape to load the computer. Well, the code got contaminated between those two operation phases. But there was never any internal check to make sure that these two files were the same. We put in such a check so that when you write the computer tape, you read it back and see if it's the same as your listing and the original file. I'll tell you, confidence in digital fly-by-wire went to zero point zero here at Dryden. We had been promoting it pretty strongly. Software never can change, it never degrades; it can never wear out. All of a sudden, you've got the wrong numbers in the computer? Even our crew chief said: "How do I know what's in that box? I want to know what's in that box." I said: "I don't have time to explain that right now." You know, he was right. I wasn't sure what was in the box. Confidence went to zero.

The first time we powered up, the system was locked into the backup system, and we could not transfer control to the Apollo computer system. We traced it to a wiring error in the airplane. We made a wiring mistake, despite our best attempts. It was due in part to the interface between our pin designations and the Draper/AC Spark Plug pin

[72]Draper Labs built the Apollo Guidance Computers for NASA under contract.

designations. In doing so, we destroyed the power supply in the Apollo Display and Keyboard, the DSKY unit.

Being in the flight test business and maybe a little naive, we said: "Well, send us another one."

And they [HQ] said: "Just a minute. There are only 12 of these in the entire world and none has ever failed. You're the first one to blow one out, you know?"

"Well, that's very interesting history. Now we need another one." I am making this sound a bit more cavalier for effect. So they searched around and pulled one out of the Apollo 15 [Command Module].

Tomayko: There was some plan to be supporting the shuttle?

Szalai: The shuttle support program on the F-8 was mostly redundancy management, the sensor tolerances. How much can a rate gyro differ from others that have been built on the same assembly line, in typical shuttle maneuvers.

Tomayko: So you were actually first, even though the shuttle was doing the research earlier?

Szalai: Right, we actually selected it first. They were looking, really doing market research, even before we started.

Tomayko: You wound up with serial numbers 1 through 9.

Szalai: It's really the good news/bad news. The machines were kind of at the end of an era in flight computers. Technology really changed after the AP-101. I think this will be the mantra for anyone buying computers for a long time.

But the AP-101s were the first flight qualified "IBM 360s." They were remarkable machines, and they enabled the F-8 Digital Fly-By-Wire to be successful because of what we could do programming them, even though they had serious hardware problems. In retrospect, it all worked out, but we would not have said this at the time. We were very unhappy people at the time the computers were failing; very, very unhappy.

My hair turned gray at that time. These were troubling times. But, looking back on it, the computer failures that occurred on the ground and in the air proved the fault tolerance technology with the real computer failures.

Tomayko: That's interesting because in-flight reports for Phase II, the computer fails on flight 2. Nothing bad happens.

Szalai: That was the most remarkable thing. Even though the rest of us were tearing our hair out, the pilot [Gary Krier] reported: "A light went on. I didn't feel anything; it kept flying." That was the proof of the pudding.

We also had had a lot of computer failures on the ground. We picked our best three computers for that first flight and didn't have a failure. But, by the time flight 2 came around, it just caught up with us. We just couldn't keep three computers running anymore. And you probably know the story of the AP-101 computer failure. It's kind of an interesting one in itself on thermal coating.

Tomayko: How did you see it?

Szalai: Well, what happened was, these circuit boards are actually multi-layered so if you want a ground, you drill a hole, and you coat it with solder on the inside, and that's a ground. So in any layer, when you design a printed circuit board, if you want to have a ground, you just bring it to that point. Then you do the same thing for 15 volt or 5 volt sources. These were called barrels. Coated cylinders. Every layer can pick up the power or ground.

On the top surface were all the parts, all the chips, and so forth: you put a protective coating over the board, which goes on like thick molasses. You have to paint it on. It's real thick. It protects the card from external contamination. If there's a floating solder ball, it can't touch any metal.

Some of the people on the assembly line said: "You know, we could save time by thinning this stuff out." So they put it into an engineering suggestion. It was evaluated and approved. Instead of taking an hour, it took five minutes to put this stuff on. Terrific approach, you know? The problem is, this stuff would seep over the edges, because now it's a thinner liquid and it would seep in between the layers of the multi-layer circuit board.

As it heated up, the coating had a coefficient of thermal expansion that was bigger than the rest of the device. It would actually force these layers apart. Now, you had this hole that was drilled through that no longer connected the two layers. The card failed. Eventually, everything kind of warmed up, reached steady state, and would start working after a while. So we had transients that occurred at the turn-on.

At the same time, the shuttle program had the computer in test at [Rockwell facility in] Downey, CA. I was talking constantly with them. They said, "We're not having any problems down here." I said, "I can't believe this." We went down there and saw they had the machines with 40-degree air coming in all the time, never turned them off so they had no thermal transients.

I told my friend in the shuttle program: "You may be in trouble. When you fire up in Palmdale I'll bet you that all the computers do not come up." And they didn't. Only three of the five came up in Palmdale. They had exactly the same problem.

Tomayko: Did you go over and volunteer to help solve the PIO problem with them?

Szalai: The F-8 wasn't an early player in that program, as it turns out. The ALT free flight no. 5 to the runway with Fred Haise at the control encountered PIO in pitch and roll. I was very alarmed at what I saw: the orbiter bounce off the runway, followed by a big lateral motion. I did not see the pitch oscillations from my vantage point. It is very hard to see a slow pitch oscillation. You have to be far away and almost dead on to see the pitch motions. But I saw the lateral motions, and I saw it cruise by still in the air. It was quite a ways down the runway before he touched down, and I knew there was a problem. You could tell he was going to stop on the runway [but] I'm not sure it was that clear to the pilots.

The problem was solved by a small group here that included John Edwards and John Smith. Smith is a patent holder on the PIO-suppression filter. It still flies in the shuttle today.

This little group of people here solved the problem. They had a lot of experience. You could not redesign the whole shuttle system at that point: they'd just finished the last major flight control hardware tests on the shuttle. No one wanted a fix that would take hundreds of thousands of hours of vehicle verification and validation. If you start playing around with the closed-loop controls it would cost a lot of money and cause a big upset in the program. The people here recognized that immediately.

So, they tried to do something outside of the closed-loop: they built this special filter. Basically, it detects the onset of the pilot-induced oscillations and reduces the pilot's stick gain automatically. It detects PIO onset by seeing a frequency component where the pilot-induced oscillation occurs. If it detects that frequency, it starts removing the

Dryden's F-8 Digiital Fly-By-Wire sits in front of the SCA and *Enterprise*. The F-8 DFBW was instrumental in solving the shuttle's latent PIO tendencies. NASA EC78-9354

pilot from the loop by reducing the pilot's gain. And it works quite well.

That was done; it was verified on various simulations, put into the shuttle system. Over the next three years, the shuttle program continued to look at other alternatives, other options. During that time, we embarked on the program with the F-8 to look at it, as well. There were several issues that had been identified as the problem with the shuttle, like excess time delay.

The F-8 did a definitive test under Don Berry, which was the first flight test with true time delay. Calspan had previously done tests using an airborne simulator with time delay simulated with analog filters. We implemented true transport time delay. We had 50, 100, 200, 300 milliseconds pure delay in the system and flew three approaches to the runway. The F-8 was one of the airplanes we used to explore the effectiveness of the filter. It was done before the first orbiter flight.

Wilt Lock said--as we always did for 25 years previously--"build breakout boxes." In other words, you have a cable; it connects two things together. Well, you build another cable where each of the wires is terminated in a box where you can kind of look at what's going on in that cable on the fly. When we first brought this up, the Apollo people and the people from the space industry said: "You don't need any of this stuff." It turns out we needed it all. I mean, if you don't have a billion dollars, you've got to come up with some clever ways of doing testing and diagnosis. So the airplane people really did come up with some really good, old-fashioned ways of ensuring the systems were working.

It was about the time that I was moving off the program that I coined the word "REBUS," Resident Backup Software System. And it turned out, the team behind me actually developed it and took it to flight. I should have patented it, probably, because it ended up in other airplanes. [With humor] My lost patent! But it actually worked; we actually induced software errors into the system, software faults that would be undetectable by any other means--probably still undetectable. And, lo and behold, it actually worked.

Tomayko: How would you rate the F-8 program from a technology transfer standpoint?

Szalai: I think it was extraordinarily effective at technology transfer.

Shot from the back seat of an inverted F-18, Dryden's photographer Carla Thomas caught *Endeavour* and the SCA over the undulating terrain near Edwards AFB as the pair headed to the Kennedy Space Center following another landing at Dryden. NASA ED08-0306-131

Documentary Histories

"Exploring The Unknown"

Logsdon, John M., ed., with Linda J. Lear, Jannelle Warren Findley, Ray A. Williamson, and Dwayne A. Day. *Exploring the Unknown: Selected Documents in the History of the U.S. Civil Space Program, Volume I, Organizing for Exploration.* NASA SP-4407, 1995.

Logsdon, John M., ed, with Dwayne A. Day, and Roger D. Launius. *Exploring the Unknown: Selected Documents in the History of the U.S. Civil Space Program, Volume II, External Relationships.* NASA SP-4407, 1996.

Logsdon, John M., ed., with Roger D. Launius, David H. Onkst, and Stephen J. Garber. *Exploring the Unknown: Selected Documents in the History of the U.S. Civil Space Program, Volume III, Using Space.* NASA SP-4407, 1998.

Logsdon, John M., ed., with Ray A. Williamson, Roger D. Launius, Russell J. Acker, Stephen J. Garber, and Jonathan L. Friedman. *Exploring the Unknown: Selected Documents in the History of the U.S. Civil Space Program, Volume IV, Accessing Space.* NASA SP-4407, 1999.

Logsdon, John M., ed., with Amy Paige Snyder, Roger D. Launius, Stephen J. Garber, and Regan Anne Newport. *Exploring the Unknown: Selected Documents in the History of the U.S. Civil Space Program, Volume V, Exploring the Cosmos.* NASA SP-4407, 2001.

Logsdon, John M., ed., with Stephen J. Garber, Roger D. Launius, and Ray A. Williamson. *Exploring the Unknown: Selected Documents in the History of the U.S. Civil Space Program, Volume VI: Space and Earth Science.* NASA SP-2004-4407, 2004.

Logsdon, John M., ed., with Roger D. Launius. *Exploring the Unknown: Selected Documents in the History of the U.S. Civil Space Program, Volume VII: Human Spaceflight: Project Mercury, Gemini, and Apollo.* NASA SP-2008-4407, 2008.

"The Wind and Beyond"

Hansen, James R., ed. *The Wind and Beyond: Journey into the History of Aerodynamics in America, Volume 1, The Ascent of the Airplane.* NASA SP-2003-4409, 2003.

Hansen, James R., ed. *The Wind and Beyond: Journey into the History of Aerodynamics in America, Volume 2, Reinventing the Airplane.* NASA SP-2007-4409, 2007.
Brief Histories of NASA

Anderson, Frank W., Jr. *Orders of Magnitude: A History of NACA and NASA, 1915-1980.* NASA SP-4403, 1981. Out of print.

Bilstein, Roger E. *Orders of Magnitude: A History of the NACA and NASA, 1915-1990.* NASA SP-4406, 1989. Out of print.

Bilstein, Roger E. *Testing Aircraft, Exploring Space: An Illustrated History of NACA and NASA.* Baltimore: Johns Hopkins University Press, 2003.

Critical Issues in the History of Spaceflight

Dick, Steven J. and Launius, Roger D., ed. *Critical Issues in the History of Spaceflight.* (NASA SP-2006-4702).

Societal Impact of Spaceflight

Dick, Steven J. and Launius, Roger D., ed. *Societal Impact of Spaceflight*. (NASA SP-2007-4801).

Dick, Steven J. and Lupisella, Mark L., ed. *Cosmos & Culture: Cultural Evolution in a Cosmic Context*. (NASA SP-2009-4802).

Memoirs:

Chertok, Boris. *Rockets and People, Volume 1*. (NASA SP-2005-4110).

Chertok, Boris. *Rockets and People: Creating a Rocket Industry, Volume II*. (NASA SP-2006-4110).

Chertok, Boris. *Rockets and People: Hot Days of the Cold War, Volume III*. (NASA SP-2009-4110).

Chertok, Boris. *Rockets and People: The Moon Race, Volume IV*. (NASA SP-2011-4110).

Mudgway, Douglas J. *William H. Pickering: America's Deep Space Pioneer*. (NASA SP-2008-4113).

NASA Historical Data Books:

Van Nimmen, Jane, and Leonard C. Bruno, with Robert L. Rosholt. *NASA Historical Data Book, Vol. I: NASA Resources, 1958-1968*. NASA SP-4012, 1976, rep. ed. 1988. Out of print.

Ezell, Linda Neuman. *NASA Historical Data Book, Vol II: Programs and Projects, 1958-1968*. NASA SP-4012, 1988. Out of print.

Ezell, Linda Neuman. *NASA Historical Data Book, Vol. III: Programs and Projects, 1969-1978*. NASA SP-4012, 1988. Out of print.

Gawdiak, Ihor, with Helen Fedor. *NASA Historical Data Book, Vol. IV: NASA Resources, 1969-1978*. NASA SP-4012, 1994.

Rumerman, Judy A. *NASA Historical Data Book, Vol. V: NASA Launch Systems, Space Transportation, Human Spaceflight, and Space Science, 1979-1988*. NASA SP-4012, 1999.

Rumerman, Judy A. *NASA Historical Data Book, Vol. VI: NASA Space Applications, Aeronautics and Space Research and Technology, Tracking and Data Acquisition/Support Operations, Commercial Programs, and Resources, 1979-1988*. NASA SP-4012, 1999. Out of print.

Rumerman, Judy A. *NASA Historical Data Book, Vol. VII: NASA Launch Systems, Space Transportation, Human Spaceflight, and Space Science, 1989-1998*. NASA SP-4012, 2009.

Rumerman, Judy A. NASA *Historical Data Book, Vol. VIII: NASA Earth Science and Space Applications, Aeronautics, Technology, and Exploration, Tracking and Data Acquisition/Space Operations, Facilities and Resources 1989–1998*. NASA SP-2012-4012, 2012. Available online only.

Astronautics and Aeronautics Chronology:

Eugene M. Emme, comp. *Aeronautics and Astronautics Chronology, 1915-1960*. Aeronautics and Astronautics: An American Chronology of Science and Technology in the Exploration of Space, 1915-1960 (Washington, DC: National Aeronautics and Space Administration, 1961). Out of print.

Eugene M. Emme, comp. *Aeronautical and Astronautical Events of 1961*. Report of the National Aeronautics and Space Administration to the Committee on Science and Astronautics, U.S. House of Representatives, 87th Cong., 2d. Sess. (Washington, DC: U.S. Government Printing Office, 1962). Out of print.

Astronautical and Aeronautical Events of 1962. Report to the Committee on Science and Astronautics, Report to the Committee on Science and Astronautics, U.S. House of Representatives, Eighty-eighth Congress, first session (Washington, DC: U.S. Government Printing Office, 1963). Out of print.

Astronautics and Aeronautics, 1963: Chronology of Science, Technology,and Policy. NASA SP-4004, 1964. Out of print.

Astronautics and Aeronautics, 1964: Chronology of Science, Technology,and Policy. NASA SP-4005, 1965. Out of print.

Astronautics and Aeronautics, 1965: Chronology of Science, Technology,and Policy. NASA SP-4006, 1966. Out of print.

Astronautics and Aeronautics, 1966: Chronology of Science, Technology,and Policy. NASA SP-4007, 1967. Out of print.

Astronautics and Aeronautics, 1967: Chronology of Science, Technology, and Policy. NASA SP-4008, 1968. Out of print.

Astronautics and Aeronautics, 1968: Chronology of Science, Technology, and Policy. NASA SP-4010, 1969. Out of print.

Astronautics and Aeronautics, 1969: Chronology of Science, Technology, and Policy. NASA SP-4014, 1970. Out of print.

Astronautics and Aeronautics, 1970: Chronology of Science, Technology, and Policy. NASA SP-4015, 1972. Out of print.

Astronautics and Aeronautics, 1971: Chronology of Science, Technology, and Policy. NASA SP-4016, 1972. Out of print.

Astronautics and Aeronautics, 1972: Chronology of Science, Technology, and Policy. NASA SP-4017, 1974. Out of print.

Astronautics and Aeronautics, 1973: Chronology of Science, Technology, and Policy. NASA SP-4018, 1975.

Astronautics and Aeronautics, 1974: Chronology of Science, Technology, and Policy. NASA SP-4019, 1977. Out of print.

Astronautics and Aeronautics, 1975: Chronology of Science, Technology, and Policy. NASA SP-4020, 1979. Out of print.

Astronautics and Aeronautics, 1976: Chronology of Science, Technology, and Policy. NASA SP-4021, 1984.

Astronautics and Aeronautics, 1977: Chronology of Science, Technology, and Policy. NASA SP-4022, 1986.

Astronautics and Aeronautics, 1978: Chronology of Science, Technology, and Policy. NASA SP-4023, 1986.

Astronautics and Aeronautics, 1979-1984: Chronology of Science, Technology, and Policy. NASA SP-4024, 1988.

Astronautics and Aeronautics, 1985: Chronology of Science, Technology, and Policy. NASA SP-4025, 1990.

Gawdiak, Ihor Y., Ramon J. Miro, and Sam Stueland, comps. *Astronautics and Aeronautics, 1986-1990: A Chronology*. NASA SP-4027, 1997.

Gawdiak, Ihor Y. and Shetland, Charles. *Astronautics and Aeronautics, 1991-1995: A Chronology*. NASA SP-2000-4028, 2000.

Lewis, Marieke and Ryan Swanson, comps. *Astronautics and Aeronautics: A Chronology, 1996-2000*. NASA SP-2009-4030, 2009. Available online only.

Ivey, William Noel and Marieke Lewis, comps. *Aeronautics and Astronautics: A Chronology, 2001-2005*. NASA SP-2010-4031, 2010. Available online only.

Buchalter, Alice R. and William Noel Ivey, comps. *Aeronautics and Astronautics: A Chronology, 2006*. NASA SP-2011-4032, 2010. Available online only.

Lewis, Marieke, comp. *Aeronautics and Astronautics: A Chronology, 2007*. NASA SP-2011-4033, 2011. Available online only.

Lewis, Marieke, comp. *Aeronautics and Astronautics: A Chronology, 2008*. NASA SP-2012-4034, 2012. Available online only.

Lewis, Marieke, comp. *Aeronautics and Astronautics: A Chronology, 2009*. NASA SP-2012-4035, 2012. Available online only.

NASA History Publications Arranged by Special Publication (SP) Nmbers

Reference Works, (SP-4000 Series)
Grimwood, James M. *Project Mercury: A Chronology*. NASA SP-4001, 1963. Out of print.

Grimwood, James M., and Barton C. Hacker, with Peter J. Vorzimmer. *Project Gemini Technology and Operations: A Chronology*. NASA SP-4002, 1969. Out of print.

Link, Mae Mills. *Space Medicine in Project Mercury*. NASA SP-4003, 1965. Out of print.

Astronautics and Aeronautics, 1963: Chronology of Science, Technology,and Policy. NASA SP-4004, 1964. Out of print.

Astronautics and Aeronautics, 1964: Chronology of Science, Technology,and Policy. NASA SP-4005, 1965. Out of print.

Astronautics and Aeronautics, 1965: Chronology of Science, Technology,and Policy. NASA SP-4006, 1966. Out of print.

Astronautics and Aeronautics, 1966: Chronology of Science, Technology,and Policy. NASA SP-4007, 1967. Out of print.

Astronautics and Aeronautics, 1967: Chronology of Science, Technology, and Policy. NASA SP-4008, 1968. Out of print.

Ertel, Ivan D., and Mary Louise Morse. *The Apollo Spacecraft: A Chronology, Volume I, Through November 7, 1962*. NASA SP-4009, 1969. Out of print.

Morse, Mary Louise, and Jean Kernahan Bays. *The Apollo Spacecraft: A Chronology, Volume II, November 8, 1962-September 30, 1964*. NASA SP-4009, 1973. Out of print.

Brooks, Courtney G., and Ivan D. Ertel. *The Apollo Spacecraft: A Chronology, Volume III, October 1, 1964-January 20, 1966*. NASA SP-4009, 1973. Out of print.

Ertel, Ivan D., and Roland W. Newkirk, with Courtney G. Brooks. *The Apollo Spacecraft: A Chronology, Volume IV, January 21, 1966-July 13, 1974*. NASA SP-4009, 1978. Out of print.

Astronautics and Aeronautics, 1968: Chronology of Science, Technology, and Policy. NASA SP-4010, 1969. Out of print.

Newkirk, Roland W., and Ivan D. Ertel, with Courtney G. Brooks. *Skylab: A Chronology*. NASA SP-4011, 1977.

Van Nimmen, Jane, and Leonard C. Bruno, with Robert L. Rosholt. *NASA Historical Data Book, Vol. I: NASA Resources, 1958-1968*. NASA SP-4012, 1976, rep. ed. 1988. Out of print.

Ezell, Linda Neuman. *NASA Historical Data Book, Vol II: Programs and Projects, 1958-1968*. NASA SP-4012, 1988. Out of print.

Ezell, Linda Neuman. *NASA Historical Data Book, Vol. III: Programs and Projects, 1969-1978*. NASA SP-4012, 1988. Out of print.

Gawdiak, Ihor, with Helen Fedor. *NASA Historical Data Book, Vol. IV: NASA Resources, 1969-1978*. NASA SP-4012, 1994.

Rumerman, Judy A. *NASA Historical Data Book, Vol. V: NASA Launch Systems, Space Transportation, Human Spaceflight, and Space Science, 1979-1988*. NASA SP-4012, 1999.

Rumerman, Judy A. *NASA Historical Data Book, Vol. VI: NASA Space Applications, Aeronautics and Space Research and Technology, Tracking and Data Acquisition/Support Operations, Commercial Programs, and Resources, 1979-1988*. NASA SP-4012, 1999 Out of print.

Rumerman, Judy A. *NASA Historical Data Book, Vol. VII: NASA Launch Systems, Space Transportation, Human Spaceflight, and Space Science, 1989-1998*. NASA SP-4012, 2009.

Rumerman, Judy A. *NASA Historical Data Book, Vol. VIII: NASA Earth Science and Space Applications, Aeronautics, Technology, and Exploration, Tracking and Data Acquisition/Space Operations, Facilities and Resources 1989–1998*. NASA SP-2012-4012, 2012. Available online only.

SP-4013 not published.

Astronautics and Aeronautics, 1969: Chronology of Science, Technology, and Policy. NASA SP-4014, 1970. Out of print.

Astronautics and Aeronautics, 1970: Chronology of Science, Technology, and Policy. NASA SP-4015, 1972. Out of print.

Astronautics and Aeronautics, 1971: Chronology of Science, Technology, and Policy. NASA SP-4016, 1972. Out of print.

Astronautics and Aeronautics, 1972: Chronology of Science, Technology, and Policy. NASA SP-4017, 1974. Out of print.

Astronautics and Aeronautics, 1973: Chronology of Science, Technology, and Policy. NASA SP-4018, 1975.

Astronautics and Aeronautics, 1974: Chronology of Science, Technology, and Policy. NASA SP-4019, 1977. Out of print.

Astronautics and Aeronautics, 1975: Chronology of Science, Technology, and Policy. NASA SP-4020, 1979. Out of print.

Astronautics and Aeronautics, 1976: Chronology of Science, Technology, and Policy. NASA SP-4021, 1984.

Astronautics and Aeronautics, 1977: Chronology of Science, Technology, and Policy. NASA SP-4022, 1986.

Astronautics and Aeronautics, 1978: Chronology of Science, Technology, and Policy. NASA SP-4023, 1986.

Astronautics and Aeronautics, 1979-1984: Chronology of Science, Technology, and Policy. NASA SP-4024, 1988.

Astronautics and Aeronautics, 1985: Chronology of Science, Technology, and Policy. NASA SP-4025, 1990.

Noordung, Hermann. *The Problem of Space Travel: The Rocket Motor*. Edited by Ernst Stuhlinger and J.D. Hunley, with Jennifer Garland. NASA SP-4026, 1995. Out of print

Gawdiak, Ihor Y., Ramon J. Miro, and Sam Stueland, comps. *Astronautics and Aeronautics, 1986-1990: A Chronology*. NASA SP-4027, 1997.

Gawdiak, Ihor Y. and Shetland, Charles. *Astronautics and Aeronautics, 1991-1995: A Chronology*. NASA SP-2000-4028, 2000.

Orloff, Richard W. *Apollo by the Numbers: A Statistical Reference*. NASA SP-2000-4029, 2000. The online version includes extensive made a number of corrections to the data in the hard copy edition. The online version does not include the original photos.

Lewis, Marieke and Ryan Swanson. *Astronautics and Aeronautics: A Chronology, 1996-2000*. NASA SP-2009-4030, 2009. Available online only.

Ivey, William Noel and Marieke Lewis, comps. *Aeronautics and Astronautics: A Chronology, 2001-2005*. NASA SP-2010-4031, 2010. Available online only.

Buchalter, Alice R. and William Noel Ivey, comps. *Aeronautics and Astronautics: A Chronology, 2006*. NASA SP-2011-4032, 2010. Available online only.

Lewis, Marieke, comp. *Aeronautics and Astronautics: A Chronology, 2007*. NASA SP-2011-4033, 2011. Available online only.

Lewis, Marieke, comp. *Aeronautics and Astronautics: A Chronology, 2008*. NASA SP-2012-4034, 2012. Available online only.

Lewis, Marieke, comp. *Aeronautics and Astronautics: A Chronology, 2009*. NASA SP-2012-4035, 2012. Available online only.

Management Histories, (SP-4100 Series)

Rosholt, Robert L. *An Administrative History of NASA, 1958-1963*. NASA SP-4101, 1966. Out of print.

Levine, Arnold S. *Managing NASA in the Apollo Era*. NASA SP-4102, 1982. Out of print.

Roland, Alex. *Model Research: The National Advisory Committee for Aeronautics,1915-1958*. NASA SP-4103, 1985.

Fries, Sylvia D. *NASA Engineers and the Age of Apollo*. NASA SP-4104, 1992. Out of print.

Glennan, T. Keith. *The Birth of NASA: The Diary of T. Keith Glennan*. Edited by J.D. Hunley. NASA SP-4105, 1993.

Seamans, Robert C. *Aiming at Targets: The Autobiography of Robert C. Seamans*. NASA SP-4106, 1996. Out of print.

Garber, Stephen J., editor. *Looking Backward, Looking Forward: Forty Years of Human Spaceflight Symposium*. NASA SP-2002-4107.

Mallick, Donald L. with Peter W. Merlin. *The Smell of Kerosene: A Test Pilot's Odyssey*. NASA SP-4108. Out of print.

Iliff, Kenneth W. and Curtis L. Peebles. *From Runway to Orbit: Reflections of a NASA Engineer*. NASA SP-2004-4109.

Chertok, Boris. *Rockets and People, Volume 1*. (NASA SP-2005-4110).

Chertok, Boris. *Rockets and People: Creating a Rocket Industry, Volume II*. (NASA SP-2006-4110).

Chertok, Boris. *Rockets and People: Hot Days of the Cold War, Volume III*.(NASA SP-2009-4110).

Chertok, Boris. *Rockets and People: The Moon Race, Volume IV*. (NASA SP-2011-4110).

Laufer, Alexander, Post, Todd, and Hoffman, Edward. *Shared Voyage: Learning and Unlearning from Remarkable Projects* (NASA SP-2005-4111).

Dawson, Virginia P. and Bowles, Mark D. *Realizing the Dream of Flight: Biographical Essays in Honor of the Centennial of Flight, 1903-2003*. (NASA SP-2005-4112).

Mudgway, Douglas J. *William H. Pickering: America's Deep Space Pioneer*. (NASA SP-2008-4113).

Dick, Steven J., Wright, Rebecca, and Johnson, Sandra; editors. *NASA at 50: Interviews with NASA's Senior Leadership*. (NASA SP-2012-4114).

Project Histories, (SP-4200 Series)

Swenson, Loyd S., Jr., James M. Grimwood, and Charles C. Alexander. *This New Ocean: A History of Project Mercury*. NASA SP-4201, 1966, reprinted 1999.

Green, Constance McLaughlin, and Milton Lomask. *Vanguard: A History*. NASA SP-4202, 1970; rep. ed. Smithsonian Institution Press, 1971. Out of print. This title is now available in a reprint edition from Dover Publications.

Hacker, Barton C., and James M. Grimwood. *On Shoulders of Titans: A History of Project Gemini*. NASA SP-4203, 1977, reprinted 2002.

Benson, Charles D. and William Barnaby Faherty. *Moonport: A History of Apollo Launch Facilities and Operations*. NASA SP-4204, 1978. The SP edition is Out of print, but the University Press of Florida has republished the book in two volumes, *Gateway to the Moon and Moon Launch!*.

Brooks, Courtney G., James M. Grimwood, and Loyd S. Swenson, Jr. *Chariots for Apollo: A History of Manned Lunar Spacecraft*. NASA SP- 4205, 1979. Out of print. This title is now available in a reprint edition from Dover Publications.

Bilstein, Roger E. *Stages to Saturn: A Technological History of the Apollo/Saturn Launch Vehicles*. NASA SP-4206, 1980 and 1996. This SP version is Out of print, but it has been reprinted by the University Press of Florida.

SP-4207 not published.

Compton, W. David, and Charles D. Benson. *Living and Working in Space:A History of Skylab*. NASA SP-4208, 1983.

Ezell, Edward Clinton, and Linda Neuman Ezell. *The Partnership: A History of the Apollo-Soyuz Test Project*. NASA SP-4209, 1978. Out of print. This title is now available in a reprint edition from Dover Publications.

Hall, R. Cargill. *Lunar Impact: A History of Project Ranger*. NASA SP-4210, 1977. Out of print. This title is now available in a reprint edition from Dover Publications.

Newell, Homer E. *Beyond the Atmosphere: Early Years of Space Science*. NASA SP-4211, 1980. Out of print. This title is now available in a reprint edition from Dover Publications.

Ezell, Edward Clinton, and Linda Neuman Ezell. *On Mars: Exploration of the Red Planet, 1958-1978*. NASA SP-4212, 1984. Out of print. This title is now available in a reprint edition from Dover Publications.

Pitts, John A. *The Human Factor: Biomedicine in the Manned Space Program to 1980*. NASA SP-4213, 1985. Out of print.

Compton, W. David. *Where No Man Has Gone Before: A History of Apollo Lunar Exploration Missions*. NASA SP-4214, 1989. This title is now available in a reprint edition from Dover Publications.

Naugle, John E. *First Among Equals: The Selection of NASA Space Science Experiments*. NASA SP-4215, 1991. Out of print.

Wallace, Lane E. *Airborne Trailblazer: Two Decades with NASA Langley's 737 Flying Laboratory*. NASA SP-4216, 1994.

Butrica, Andrew J. *Beyond the Ionosphere: Fifty Years of Satellite Communications*. NASA SP-4217, 1997.

Butrica, Andrew J. *To See the Unseen: A History of Planetary Radar Astronomy*. NASA SP-4218, 1996.

Mack, Pamela E., ed. *From Engineering Science to Big Science: The NACA and NASA Collier Trophy Research Project Winners*. NASA SP-4219, 1998.

Reed, R. Dale. *Wingless Flight: The Lifting Body Story*. NASA SP-4220, 1998.

Heppenheimer, T. A. *The Space Shuttle Decision: NASA's Search for a Reusable Space Vehicle.* NASA SP-4221, 1999.

Hunley, J. D., ed. *Toward Mach 2: The Douglas D-558 Program.* NASA SP-4222, 1999.

Swanson, Glen E., ed. *"Before This Decade is Out..." Personal Reflections on the Apollo Program.* NASA SP-4223, 1999.

Tomayko, James E. *Computers Take Flight: A History of NASA's Pioneering Digital Fly-By-Wire Project.* NASA SP-4224, 2000.

Morgan, Clay. *Shuttle-Mir: The United States and Russia Share History's Highest Stage.* NASA SP-2001-4225.

Leary, William M. *We Freeze to Please: A History of NASA's Icing Research Tunnel and the Quest for Safety.* NASA SP-2002-4226, 2002.

Mudgway, Douglas J. *Uplink-Downlink: A History of the Deep Space Network, 1957-1997.* NASA SP-2001-4227.

SP-4228 and SP-4229 not published.

Dawson, Virginia P. and Mark D. Bowles. *Taming Liquid Hydrogen: The Centaur Upper Stage Rocket, 1958-2002 .* NASA SP-2004-4230. Out of print.

Meltzer, Michael. *Mission to Jupiter: A History of the Galileo Project.* NASA SP-2007-4231.

Heppenheimer, T.A. *Facing the Heat Barrier: A History of Hypersonics.* NASA SP-2007-4232.

Tsiao, Sunny. *"Read You Loud and Clear!" The Story of NASA's Spaceflight Tracking and Data Network.* NASA SP-2007-4233.

Meltzer, Michael. *When Biospheres Collide: A History of NASA's Planetary Protection Programs.* NASA SP-2011-4234.

Center Histories, (SP-4300 Series)
Rosenthal, Alfred. *Venture into Space: Early Years of Goddard Space Flight Center.* NASA SP-4301, 1985. Out of print.

Hartman, Edwin, P. *Adventures in Research: A History of Ames Research Center, 1940-1965.* NASA SP-4302, 1970. Out of print.

Hallion, Richard P. *On the Frontier: Flight Research at Dryden, 1946-1981.* NASA SP-4303, 1984. Out of print.

Muenger, Elizabeth A. *Searching the Horizon: A History of Ames Research Center, 1940-1976.* NASA SP-4304, 1985.

Hansen, James R. *Engineer in Charge: A History of the Langley Aeronautical Laboratory,1917-1958.* NASA SP-4305, 1987. Out of print.

Dawson, Virginia P. *Engines and Innovation: Lewis Laboratory and American Propulsion Technology.* NASA

SP-4306, 1991. Out of print.

Dethloff, Henry C. *"Suddenly Tomorrow Came...": A History of the Johnson Space Center, 1957-1990.* NASA SP-4307, 1993. Out of print.

Hansen, James R. *Spaceflight Revolution: NASA Langley Research Center from Sputnik to Apollo.* NASA SP-4308, 1995. Out of print.

Wallace, Lane E. *Flights of Discovery: 50 Years of Flight Research at the NASA Dryden Flight Research Center.* NASA SP-4309, 1996.

Herring, Mack R. *Way Station to Space: A History of the John C. Stennis Space Center.* NASA SP-4310, 1997.

Wallace, Harold D., Jr. *Wallops Station and the Creation of an American Space Program.* NASA SP-4311, 1997.

Wallace, Lane E. *Dreams, Hopes, Realities. NASA's Goddard Space Flight Center: The First Forty Years.* NASA SP-4312, 1999. Out of print.

Dunar, Andrew J. and Waring, Stephen P. *Power to Explore: A History of Marshall Space Flight Center, 1960-1990.* NASA SP-4313, 1999. Out of print.

Bugos, Glenn E. *Atmosphere of Freedom: Sixty Years at the NASA Ames Research Center.* NASA SP-2000-4314, 2000.

Bugos, Glenn E. *Atmosphere of Freedom: Seventy Years at the NASA Ames Research Center.* NASA SP-2010-4314, 2010. This is an update of the 2000 edition.

SP-4315 not published.

Schultz, James. *Crafting Flight: Aircraft Pioneers and the Contributions of the Men and Women of NASA Langley Research Center.* NASA SP-2003-4316, 2003.

Bowles, Mark D. Science in Flux: *NASA's Nuclear Program at Plum Brook Station, 1955-2005.* NASA SP-2006-4317.

Wallace, Lane E. *Flights of Discovery: 60 Years of Flight Research at the NASA Dryden Flight Research Center.* NASA SP-4318, 2006. Revised version of SP-4309.

Arrighi, Robert S. *Revolutionary Atmosphere: The Story of the Altitude Wind Tunnel and the Space Power Chambers.* NASA SP-2010-4319, 2010.

General Histories, (SP-4400 Series)
Corliss, William R. *NASA Sounding Rockets, 1958-1968: A Historical Summary.* NASA SP-4401, 1971. Out of print.

Wells, Helen T., Susan H. Whiteley, and Carrie Karegeannes. *Origins of NASA Names.* NASA SP-4402, 1976. Out of print.

Anderson, Frank W., Jr. *Orders of Magnitude: A History of NACA and NASA, 1915-1980.* NASA SP-4403, 1981. Out of print.

Sloop, John L. *Liquid Hydrogen as a Propulsion Fuel, 1945-1959*. NASA SP-4404, 1978. Out of print.

Roland, Alex. *A Spacefaring People: Perspectives on Early Spaceflight. NASA SP-4405, 1985*. Out of print.

Bilstein, Roger E. *Orders of Magnitude: A History of the NACA and NASA, 1915-1990*. NASA SP-4406, 1989. Out of print.

Logsdon, John M., ed., with Linda J. Lear, Jannelle Warren Findley, Ray A. Williamson, and Dwayne A. Day. *Exploring the Unknown: Selected Documents in the History of the U.S. Civil Space Program, Volume I, Organizing for Exploration*. NASA SP-4407, 1995.

Logsdon, John M., ed, with Dwayne A. Day, and Roger D. Launius. *Exploring the Unknown: Selected Documents in the History of the U.S. Civil Space Program, Volume II, External Relationships*. NASA SP-4407, 1996.

Logsdon, John M., ed., with Roger D. Launius, David H. Onkst, and Stephen J. Garber. *Exploring the Unknown: Selected Documents in the History of the U.S. Civil Space Program, Volume III, Using Space*. NASA SP-4407,1998.

Logsdon, John M., ed., with Ray A. Williamson, Roger D. Launius, Russell J. Acker, Stephen J. Garber, and Jonathan L. Friedman. *Exploring the Unknown: Selected Documents in the History of the U.S. Civil Space Program,Volume IV, Accessing Space*. NASA SP-4407, 1999.

Logsdon, John M., ed., with Amy Paige Snyder, Roger D. Launius, Stephen J. Garber, and Regan Anne Newport. *Exploring the Unknown: Selected Documents in the History of the U.S. Civil Space Program, Volume V, Exploring the Cosmos*. NASA SP-4407, 2001.

Logsdon, John M., ed., with Stephen J. Garber, Roger D. Launius, and Ray A. Williamson. *Exploring the Unknown: Selected Documents in the History of the U.S. Civil Space Program, Volume VI: Space and Earth Science*. NASA SP-2004-4407, 2004.

Logsdon, John M., ed., with Roger D. Launius. *Exploring the Unknown: Selected Documents in the History of the U.S. Civil Space Program, Volume VII: Human Spaceflight: Project Mercury, Gemini, and Apollo*. NASA SP-2008-4407, 2008.

Siddiqi, Asif A., *Challenge to Apollo: The Soviet Union and the Space Race, 1945-1974*. NASA SP-2000-4408, 2000.

Hansen, James R., ed. *The Wind and Beyond: Journey into the History of Aerodynamics in America, Volume 1, The Ascent of the Airplane*. NASA SP-2003-4409, 2003.

Hansen, James R., ed. *The Wind and Beyond: Journey into the History of Aerodynamics in America, Volume 2, Reinventing the Airplane*. NASA SP-2007-4409, 2007.

Hogan, Thor. *Mars Wars: The Rise and Fall of the Space Exploration Initiative*. NASA SP-2007-4410, 2007.

Vakoch, Douglas A., ed. *Psychology of Space Exploration*. NASA SP-2011-4411, 2011.

Monographs in Aerospace History (SP-4500 Series)
Launius, Roger D. and Aaron K. Gillette, comps. Toward a History of the Space Shuttle: An Annotated Bibliography. Monographs in Aerospace History, No. 1, 1992. Out of print.

Launius, Roger D., and J.D. Hunley, comps. *An Annotated Bibliography of the Apollo Program*. Monographs in Aerospace History No. 2, 1994. Out of print

Launius, Roger D. *Apollo: A Retrospective Analysis*. Monographs in Aerospace History, No. 3, 1994.

Hansen, James R. *Enchanted Rendezvous: John C. Houbolt and the Genesis of the Lunar-Orbit Rendezvous Concept*. Monographs in Aerospace History, No. 4, 1995.

Gorn, Michael H. *Hugh L. Dryden's Career in Aviation and Space*. Monographs in Aerospace History, No. 5, 1996.

Powers, Sheryll Goecke. *Women in Flight Research at NASA Dryden Flight Research Center from 1946 to 1995*. Monographs in Aerospace History, No. 6, 1997.

Portree, David S.F. and Robert C. Trevino. *Walking to Olympus: An EVA Chronology*. Monographs in Aerospace History, No. 7, 1997. Out of print

Logsdon, John M., moderator. *Legislative Origins of the National Aeronautics and Space Act of 1958: Proceedings of an Oral History Workshop*. Monographs in Aerospace History, No. 8, 1998. Out of print. This monograph is available online in a text-only pdf file or a pdf file with graphics.

Rumerman, Judy A., comp. *U.S. Human Spaceflight, A Record of Achievement 1961-1998*. Monographs in Aerospace History, No. 9, 1998.

Portree, David S. F. *NASA's Origins and the Dawn of the Space Age*. Monographs in Aerospace History, No. 10, 1998.

Logsdon, John M. *Together in Orbit: The Origins of International Cooperation in the Space Station*. Monographs in Aerospace History, No. 11, 1998.

Phillips, W. Hewitt. *Journey in Aeronautical Research: A Career at NASA Langley Research Center*. Monographs in Aerospace History, No. 12, 1998.

Braslow, Albert L. *A History of Suction-Type Laminar-Flow Control with Emphasis on Flight Research*. Monographs in Aerospace History, No. 13, 1999.

Logsdon, John M., moderator. *Managing the Moon Program: Lessons Learned From Apollo*. Monographs in Aerospace History, No. 14, 1999.

Perminov, V.G. *The Difficult Road to Mars: A Brief History of Mars Exploration in the Soviet Union*. Monographs in Aerospace History, No. 15, 1999. Out of print. This book is available in pdf.

Tucker, Tom. Touchdown: *The Development of Propulsion Controlled Aircraft at NASA Dryden*. Monographs in Aerospace History, No. 16, 1999.

Maisel, Martin, Giulanetti, Demo J., and Dugan, Daniel C. *The History of the XV-15 Tilt Rotor Research Aircraft: From Concept to Flight*. Monographs in Aerospace History, No. 17, 2000 (NASA SP-2000-4517).

Jenkins, Dennis R. *Hypersonics Before the Shuttle: A Concise History of the X-15 Research Airplane*. Monographs in Aerospace History, No. 18, 2000 (NASA SP-2000-4518). Out of print

Chambers, Joseph R. *Partners in Freedom: Contributions of the Langley Research Center to U.S. Military*

Aircraft of the 1990s. Monographs in Aerospace History, No. 19, 2000 (NASA SP-2000-4519).

Waltman, Gene L. *Black Magic and Gremlins: Analog Flight Simulations at NASA's Flight Research Center*. Monographs in Aerospace History, No. 20, 2000 (NASA SP-2000-4520).

Portree, David S.F. *Humans to Mars: Fifty Years of Mission Planning, 1950-2000*. Monographs in Aerospace History, No. 21, 2001 (NASA SP-2001-4521).

Thompson, Milton O. with J.D. Hunley. *Flight Research: Problems Encountered and What they Should Teach Us*. Monographs in Aerospace History, No. 22, 2001 (NASA SP-2001-4522).

Tucker, Tom. *The Eclipse Project*. Monographs in Aerospace History, No. 23, 2001 (NASA SP-2001-4523).

Siddiqi, Asif A. *Deep Space Chronicle: A Chronology of Deep Space and Planetary Probes 1958-2000*. Monographs in Aerospace History, No. 24, 2002 (NASA SP-2002-4524). Out of print

Merlin, Peter W. *Mach 3+: NASA/USAF YF-12 Flight Research, 1969-1979*. Monographs in Aerospace History, No. 25, 2001 (NASA SP-2001-4525). Out of print

Anderson, Seth B. *Memoirs of an Aeronautical Engineer: Flight Tests at Ames Research Center: 1940-1970*. Monographs in Aerospace History, No. 26, 2002 (NASA SP-2002-4526)

Renstrom, Arthur G. *Wilbur and Orville Wright: A Bibliography Commemorating the One-Hundredth Anniversary of the First Powered Flight on December 17, 1903*. Monographs in Aerospace History, No. 27, 2002 (NASA SP-2002-4527).

Monograph 28 (NASA SP-4528) not published.

Chambers, Joseph R. *Concept to Reality: Contributions of the NASA Langley Research Center to U.S. Civil Aircraft of the 1990s*. Monographs in Aerospace History, No. 29, 2003. (SP-2003-4529).

Peebles, Curtis, editor. *The Spoken Word: Recollections of Dryden History, The Early Years*. Monographs in Aerospace History, No. 30, 2003. (SP-2003-4530).

Jenkins, Dennis R.; Landis, Tony; and Miller, Jay. *American X-Vehicles: An Inventory- X-1 to X-50*. Monographs in Aerospace History, No. 31, 2003 (SP-2003-4531). Out of print

Renstrom, Arthur G. *Wilbur and Orville Wright: A Chronology Commemorating the One-Hundredth Anniversary of the First Powered Flight on December 17, 1903*. Monographs in Aerospace History, No. 32, 2003. (NASA SP-2003-4532).

Bowles, Mark D. and Arrighi, Robert S. *NASA's Nuclear Frontier: The Plum Brook Research Reactor*. Monographs in Aerospace History, No. 33, 2004. (SP-2004-4533).

Wallace, Lane and Christian Gelzer. *Nose Up: High Angle-of-Attack and Thrust Vectoring Research at NASA Dryden 1979-2001*. Monographs in Aerospace History, No. 34, 2009. (NASA SP-2009-4534).

Matranga, Gene J.; Ottinger, C. Wayne; Jarvis, Calvin R.; and Gelzer, D. Christian. *Unconventional, Contrary, and Ugly: The Lunar Landing Research Vehicle*. Monographs in Aerospace History, No. 35, 2006. (NASA SP-2004-4535). Out of print

McCurdy, Howard E. *Low Cost Innovation in Spaceflight: The History of the Near Earth Asteroid Rendezvous*

(NEAR) Mission. Monographs in Aerospace History, No. 36, 2005. (NASA SP-2005-4536).

Seamans, Robert C. Jr. *Project Apollo: The Tough Decisions*. Monographs in Aerospace History, No. 37, 2005. (NASA SP-2005-4537).

Lambright, W. Henry. *NASA and the Environment: The Case of Ozone Depletion*. Monographs in Aerospace History, No. 38, 2005. (NASA SP-2005-4538).

Chambers, Joseph R. *Innovation in Flight: Research of the NASA Langley Research Center on Revolutionary Advanced Concepts for Aeronautics*. Monographs in Aerospace History, No. 39, 2005. (NASA SP-2005-4539). This monograph is only available on-line.

Phillips, W. Hewitt. *Journey Into Space Research: Continuation of a Career at NASA Langley Research Center*. Monographs in Aerospace History, No. 40, 2005. (NASA SP-2005-4540). This monograph is only available on-line.

Rumerman, Judy A., comp. *U.S. Human Spaceflight: A Record of Achievement, 1961-2006*. Monographs in Aerospace History No. 41, 2007. (NASA SP-2007-4541).

Peebles, Curtis. *The Spoken Word II: Recollections of Dryden History, Beyond the Sky*. Monographs in Aerospace History, No. 42, 2011. (NASA SP-2011-4542).

Dick, Steven J.; Garber, Stephen J.; and Odom, Jane H., comp. *Research in NASA History. A Guide to the NASA History Program*. Third Edition. Monographs in Aerospace History, No. 43, 2009. (NASA SP-2009-4543).

Merlin, Peter W. Ikhana: *Unmanned Aircraft System Western States Fire Missions*. Monographs in Aerospace History, No. 44, 2009. (NASA SP-2009-4544).

Fisher, Steven C. and Rahman, Shamim A., eds. *Remembering the Giants: Apollo Rocket Propulsion Development*. Monographs in Aerospace History, No. 45, 2009. (NASA SP-2009-4545).

Gelzer, Christian. *Fairing Well: From Shoebox to Bat Truck and Beyond, Aerodynamic Truck Research at NASA's Dryden Flight Research Center*. Monographs in Aerospace History, No. 46, 2011. (NASA SP-2011-4546).
Arrighi, Robert. *Pursuit of Power: NASA's Propulsion Systems Laboratory No. 1 and 2*. Monographs in Aerospace History, No. 48, 2012. (NASA SP-2012-4548).

Goodrich, Malinda K.; Buchalter, Alice R.; and Miller, Patrick M. of the Federal Research Division, Library of Congress, comp. *Toward A History of the Space Shuttle, An Annotated Bibliography, Part 2 (1992–2011)* Monographs in Aerospace History, No. 49, 2012. (NASA SP-2012-4549).

Electronic Media (SP-4600 Series)
Remembering Apollo 11: The 30th Anniversary Data Archive CD-ROM. (NASA SP-4601, 1999) Out of print.

Remembering Apollo 11: The 35th Anniversary Data Archive CD-ROM. (NASA SP-2004-4601, 2004). This is an update of the 1999 edition.

The Mission Transcript Collection: U.S. Human Spaceflight Missions from Mercury Redstone 3 to Apollo 17. CD-ROM (SP-2000-4602, 2001). Out of print. Now available commerically from CG Publishing.

Shuttle-Mir: the United States and Russia Share History's Highest Stage. (NASA SP-2001-4603, 2002). This

CD-ROM is also available from NASA CORE.

U.S. Centennial of Flight Commission presents Born of Dreams ~ Inspired by Freedom. CD-ROM (NASA SP-2004-4604, 2004).

Of Ashes and Atoms: A Documentary on the NASA Plum Brook Reactor Facility. DVD (NASA SP-2005-4605).

Taming Liquid Hydrogen: The Centaur Upper Stage Rocket Interactive CD-ROM. CD-ROM (NASA SP-2004-4606, 2004).

Fueling Space Exploration: The History of NASA's Rocket Engine Test Facility DVD. DVD (NASA SP-2005-4607).

Altitude Wind Tunnel at NASA Glenn Research Center. An Interactive History. CD-ROM (NASA SP-2008-4608).

A Tunnel Through Time: The History of NASA's Altitude Wind Tunnel. DVD (NASA SP-2010-4609).
Conference Proceedings (SP-4700 Series)

Conference Proceedings (SP-4700 Series)
Dick, Steven J. and Cowing, Keith L, ed. *Risk and Exploration: Earth, Sea and the Stars.* (NASA SP-2005-4701).

Dick, Steven J. and Launius, Roger D., ed. *Critical Issues in the History of Spaceflight.* (NASA SP-2006-4702).

Dick, Steven J., ed. *Remembering the Space Age.* (NASA SP-2008-4703).

Dick, Steven J., ed. *NASA 50th Anniversary Proceedings. NASA's First 50 Years: Historical Perspectives.* (NASA SP-2010-4704).

Societal Impact of Spaceflight (SP-4800 Series)
Dick, Steven J. and Launius, Roger D. *Societal Impact of Spaceflight.* (NASA SP-2007-4801).

Dick, Steven J. and Lupisella, Mark L., ed. *Cosmos & Culture: Cultural Evolution in a Cosmic Context.* (NASA SP-2009-4802).

Historical Reports (HHR Series)
Boone, W. Fred. *NASA Office of Defense Affairs: The First Five Years.* (NASA HHR-32, 1970).
Conference Proceedings (CPs)

Life in the Universe: Proceedings of a conference held at NASA Ames Research Center Moffet Field, California, June 19-20, 1979. (NASA CP-2156, 1981), edited by John Billingham.

Proceedings of the X-15 First Flight 30th Anniversary Celebration of June 8, 1989. (NASA CP-3105, 1991).
Contractor Reports (CRs)

Computers in Spaceflight: The NASA Experience. By James E. Tomayko. (NASA Contractor Report 182505, 1988

Educational Publications (EPs)

Apollo 13 "Houston, we've got a problem." (NASA EP-76, 1970). Out of print.

On the Moon with Apollo 16: A Guide to the Descartes Region. (NASA EP-95, 1972)

Skylab: A Guidebook. (NASA EP-107, 1973), by Leland F. Belew and Ernst Stuhlinger.

Why Man Explores. (NASA EP-125, 1976).

Spacelab: An International Short-Stay Orbiting Laboratory. (NASA EP-165) by Walter Froehlich.

A Meeting with the Universe: Science Discoveries from the Space Program. (NASA EP-177, 1981).
NASA Publications (NPs)

Science in Orbit: The Shuttle & Spacelab Experience: 1981-1986. (NASA NP-119, Marshall Space Flight Center, 1988).

Special Publications (SPs)
Results of the Second Manned Suborbital Space Flight, July 21, 1961. (Pre-SP, NASA, 1961).

Results of the Second U.S. Manned Orbital Space Flight. (NASA SP-6, 1962).

Results of the Third U.S. Manned Orbital Space Flight. (NASA SP-12, 1962).

Mercury Project Summary including Results of the Fourth Manned Orbital Flight. (NASA SP-45, 1963).

X-15 Research Results With a Selected Bibliography. (NASA SP-60, 1965).

Exploring Space with a Camera (NASA SP-168, 1968).

Aerospace Food Technology (NASA SP-202, 1969).

What Made Apollo a Success? (NASA SP-287, 1971).

Evolution of the Solar System. (NASA SP-345, 1976).

Pioneer Odyssey (NASA SP-349/396, revised edition, 1977) by Richard Fimmel, William Swindell, and Eric Burgess.

Apollo Expeditions to the Moon. (NASA SP-350, 1975). Out of print. This title is now available in a reprint edition from Dover Publications.

Apollo Over the Moon: A View From Orbit. (NASA SP-362, 1978) edited by Harold Masursky, G.W. Colton, and Farouk El-Baz.

Introduction to the Aerodynamics of Flight. (NASA SP-367, 1975) by Theodore A. Talay.

Biomedical Results of Apollo. (NASA SP-368, 1975), edited by Richard S. Johnston, Lawrence F. Dietlein, M.D., and Charles A. Berry, M.D.

Skylab EREP Investigations Summary. (NASA SP-399, 1978).

Skylab: Our First Space Station. (NASA SP-400, 1977), edited by Leland F. Belew.

Skylab, Classroom in Space. (NASA SP-401, 1977), edited by Lee Summerlin.

A New Sun: Solar Results from Skylab. (NASA SP-402, 1979) by John A. Eddy and edited by Rein Ise.

Skylab's Astronomy and Space Sciences. (NASA SP-404, 1979), edited by Charles A. Lundquist.

The Space Shuttle. (NASA SP-407, 1976).

The Search For Extraterrestrial Intelligence. (NASA SP-419, 1977), edited by Philip Morrison, John Billingham, and John Wolfe.

Atlas of Mercury. (NASA SP-423, 1978) by Merton E. Davies, Stephen E. Dwornik, et. al.

The Voyage of Mariner 10: Mission to Venus and Mercury. (NASA SP-424, 1978) by James A. Dunne and Eric Burgess.

The Martian Landscape. (NASA SP-425, 1978).

The Space Shuttle at Work. (NASA SP-432/EP-156 1979) by Howard Allaway.

Project Orion: A Design Study of a System for Detecting Extrasolar Planets. (NASA SP-436, 1980), edited by David C. Black.

Wind Tunnels of NASA. (NASA SP-440, 1981).

Viking Orbiter Views of Mars. (NASA SP-441, 1980).

The High Speed Frontier: Case Histories of Four NACA Programs, 1920-1950. (NASA SP-445, 1980.)

The Star Splitters: The High Energy Astronomy Observatories. (NASA SP-466, 1984) by Wallace H. Tucker.

Planetary Geology in the 1980s. (NASA SP-467, 1985) by Joseph Veverka.

Quest for Performance: The Evolution of Modern Aircraft. (NASA SP-468, 1985).

The Long Duration Exposure Facility (LDEF): Mission 1 Experiments. (SP-473, 1984) ed. by Lenwood G. Clark, William H. Kinar, et. al.

Voyager 1 and 2, Atlas of Saturnian Satellites. (NASA SP-474, 1984) edited by Raymond Batson.

Far Travelers: The Exploring Machines. (NASA SP-480, 1985) by Oran W. Nicks.

The Impact of Science on Society. (NASA SP-482, 1985) by James Burke, Jules Bergman, and Isaac Asimov.

Living Aloft: Human Requirements for Extended Spaceflight. (NASA SP-483, 1985).

Space Shuttle Avionics System. (NASA SP-504, 1989) by John F. Hanaway and Robert W. Moorehead.

Life Into Space: Space Life Sciences Research, Volumes I - III. 1965-2003 (NASA SP-534).

Flight Research at Ames, 1940-1997. (NASA SP-3300, 1998).

Unmanned Space Project Management: Surveyor and Lunar Orbiter. (NASA SP-4901, 1972) by Erasmus H. Kloman.

The Planetary Quarantine Program. (NASA SP-4902, 1974).

Constellation Program Lessons Learned Volume 1: Executive Summary. (NASA SP-6127, 2011)

Spaceborne Digital Computer Systems. (NASA SP-8070, 1971).

Celebrating a Century of Flight. (NASA SP-2002-09-511-HQ). Edited by Tony Springer.

Wings in Orbit: Scientific and Engineering Legacies of the Space Shuttle. (NASA SP-2010-3409). Executive Editor: Wayne Hale.

Technical Memoranda and Technical Notes (TMs and TNs)
Present and Future State of the Art in Guidance Computer Memories. (NASA TN D-4224, 1967) by Robert C. Ricci.

Other Government Publications Related to Aerospace History
History of Research in Space Biology and Biodynamics at the Air Force Missile Development Center, Holloman Air Force Base, New Mexico, 1946-1958.

Report of the Apollo 13 Review Board. (a.k.a. the Cortright Commission).

Report of the Presidential Commission on the Space Shuttle Challenger Accident. (commonly called the Rogers Commission Report), June 1986 and Implementations of the Recommendations, June 1987

Transiting from Air to Space: The North American X-15. This case study by Robert S. Houston, Richard P. Hallion, and Ronald G. Boston is a long chapter in *The Hypersonic Revolution: Case Studies in the History of Hypersonic Technology* (Air Force History and Museums Program: 1998).

Space Handbook: Astronautics and its Applications. This 1959 publication was a staff report of the Congressional Select Committee on Astronautics and Space Exploration.

The First Century of Flight: NACA/NASA Contributions to Aeronautics. This is an informative and attractive Web exhibit set up in a timeline format.

Space Station Requirements and Transportation Options for Lunar Outpost. Proceedings of the Twenty-Seventh Space Congress, April 24-27, 1990, Cocoa Beach, Florida. NASA Office of Logic Design (NASA, 1990).

Space Station Freedom Accommodation of the Human Exploration Initiative. Proceedings of the Twenty-Seventh Space Congress, April 24-27, 1990, Cocoa Beach, Florida. NASA Office of Logic Design (NASA, 1990).

NASA History Titles Published by NASA Centers
Tomayko, James E., author, and Christian Gelzer, editor. *The Story of Self-Repairing Flight Control Systems*. Dryden Historical Study #1.

Dawson, Virginia. *Ideas Into Hardware: A History of the Rocket Engine Test Facility at the NASA Glenn Research Center*. Cleveland: NASA Glenn Research Center, 2004.

Guide to Magellan Image Interpretation. (JPL-93-24) by John Ford, Jeffrey Plaut, et. al.

Magellan: The Unveiling of Venus. (JPL-400-345, 1989).

The Apollo Program Summary Report. (Document # JSC-09423, April 1975).

Saturn Illustrated Chronology. (MHR-5, Marshall Space Flight Center, fifth edition, 1971) prepared by David S. Akens.

NASA History Published by Commercial Presses

New Series in NASA History Published by the Johns Hopkins University Press:
Cooper, Henry S. F., Jr. *Before Lift-off: The Making of a Space Shuttle Crew*. Baltimore: Johns Hopkins University Press, 1987.

McCurdy, Howard E. *The Space Station Decision: Incremental Politics and Technological Choice*. Baltimore: Johns Hopkins University Press, 1990.

Hufbauer, Karl. *Exploring the Sun: Solar Science Since Galileo*. Baltimore:Johns Hopkins University Press, 1991.

McCurdy, Howard E. *Inside NASA: High Technology and Organizational Change in the U.S. Space Program*. Baltimore: Johns Hopkins University Press,1993.

Lambright, W. Henry. *Powering Apollo: James E. Webb of NASA*. Baltimore: Johns Hopkins University Press, 1995.

Bromberg, Joan Lisa. *NASA and the Space Industry*. Baltimore: Johns Hopkins University Press, 1999.

Beattie, Donald A. *Taking Science to the Moon: Lunar Experiments and the Apollo Program*. Baltimore: Johns Hopkins University Press, 2001.

McCurdy, Howard E. *Faster, Better, Cheaper: Low-Cost Innovation in the U.S. Space Program*. Baltimore: Johns Hopkins University Press, 2001.

Johnson, Stephen B. *The Secret of Apollo: Systems Management in American and European Space Programs*. Baltimore: Johns Hopkins University Press, 2002.

Lambright, W. Henry, editor. *Space Policy in the 21st Century*. Baltimore: Johns Hopkins University Press, 2002.

Bilstein, Roger E. *Testing Aircraft, Exploring Space: An Illustrated History of NACA and NASA*. Baltimore: Johns Hopkins University Press, 2003.

Butrica, Andrew J. *Single Stage to Orbit: Politics, Space Technology, and the Quest for Reusable Rocketry*. Baltimore: Johns Hopkins University Press, 2005.

Conway, Erik M. *High-Speed Dreams: NASA and the Technopolitics of Supersonic Transportation, 1945-1999*. Baltimore: Johns Hopkins University Press, 2005.

Launius, Roger D. and Howard E. McCurdy. *Robots in Space: Technology, Evolution, and Interplanetary Travel*. Baltimore: Johns Hopkins University Press, 2008.

Conway, Erik M. *Atmospheric Science at NASA: A History*. Baltimore: Johns Hopkins University Press, 2008.

Dickson, Paul. *A Dictionary of the Space Age*. Baltimore: Johns Hopkins University Press, 2009.

NASA History Titles Published by Texas A&M University Press
Schorn, Ronald A. *Planetary Astronomy: From Ancient Times to the Third Millennium*. College Station: Texas A&M University Press, 1998.

NASA History Titles Published by The University Press of Kentucky
Gorn, Michael H. *Expanding the Envelope: Flight Research at NACA and NASA*. Lexington: The University Press of Kentucky, 2001.

Reed, R. Dale. *Wingless Flight: The Lifting Body Story*. Lexington: The University Press of Kentucky, 2002.

Ed. by Launius, Roger D. and Dennis R. Jenkins. *To Reach the High Frontier: A History of U.S. Launch Vehicles*. Lexington: The University Press of Kentucky, 2002.

NASA History Titles Published by the University Press of Florida
Ed. by Swanson, Glen W. *"Before This Decade is Out..:" Personal Reflections on the Apollo Program*. Gainesville: The University Press of Florida, 2002.

Benson, Charles D. and William B. Faherty. *Moon Launch!: A History of the Saturn-Apollo Launch Operations*. Gainesville: The University Press of Florida, 2001.

Benson, Charles D. and William B. Faherty. *Gateway to the Moon: Building the Kennedy Space Center Launch Complex*. Gainesville: The University Press of Florida, 2001.

Bilstein, Roger E. *Stages to Saturn: A Technological History of the Apollo/Saturn Launch Vehicles*. (Originally published as NASA SP-4206 in 1980 and reprinted in 1996). Gainesville: The University Press of Florida, 2003.

Siddiqi, Asif A. *The Soviet Space Race with Apollo*. Gainesville: The University Press of Florida, 2003.

Siddiqi, Asif A. *Sputnik and the Soviet Space Challenge*. Gainesville: The University Press of Florida, 2003.

Lipartito, Kenneth and Butler, Orville R. *A History of the Kennedy Space Center*. Gainesville: The University Press of Florida, 2007.

NASA History Titles Published by Harwood Academic Press
Ed. by Roger D. Launius, John M. Logsdon and Robert W. Smith. *Reconsidering Sputnik: Forty Years Since the Soviet Satellite*. London: Harwood Academic Press, 2000.

NASA History Titles Published by the University of Illinois Press
Ed. by Roger D. Launius and Howard McCurdy. *Spaceflight and the Myth of Presidential Leadership*. Urbana, IL: University of Illinois Press, 1997.

NASA History Titles Published by Greenwood Press
Launius, Roger D. *Frontiers of Space Exploration*. Westport, CT: Greenwood Press, 1998.

NASA History Titles Published by the Smithsonian Institution Press
Heppenheimer, T.A. *Development of the Shuttle, 1972-1981*. Washington, DC: Smithsonian Institution Press, 2002.

Dethloff, Henry C. and Ronald A. Schorn. *Voyager's Grand Tour: To the Outer Planets and Beyond*. Washington, DC: Smithsonian Institution Press, 2003.

Hallion, Richard P. and Michael H. Gorn. *On the Frontier: Experimental Flight at NASA Dryden*. Washington, DC: Smithsonian Institution Press, 2003.

Launius, Roger D. and Andrew K. Johnston. *Atlas of Space Exploration*. Smithsonian Institution Press, 2009.

Neufeld, Michael J. *Spacefarers: Images of Astronauts and Cosmonauts in the Heroic Era of Spaceflight*. Smithsonian Institution Scholarly Press, 2013.

NASA History Titles Published by CG Publishing, Inc.
The Mission Transcript Collection: U.S. Human Spaceflight Missions From Mercury Redstone 3 to Apollo 17 (NASA SP-2000-4602).

NASA History Titles Published by Abrams Press
Dick, Steven, editor, et. al. *America In Space: NASA's First Fifty Years*. New York: Abrams, 2007.
NASA History Titles Published by MIT Press

Clancey, William J. *Working on Mars: Voyagers of Scientific Discovery with the Mars Exploration Rovers*. MIT Press, 2012.